Traditional Cattle Breeds
and how to keep them

Frère de Limbourg, Très Riches Heures du Duc de Berry (1415 - 1416).
March - Tilling. Musée Condé, Chantilly.

Traditional Cattle Breeds
and how to keep them

Young White Park Bulls Sparring. (Photo, Eileen Hayes°).

First published 2004

Copyright © Peter King

A catalogue record for this book is available from the British Library

Published by
Farming Books and Videos Ltd.
PO Box 536, Preston PR2 9ZY
United Kingdom

www.farmingbooksandvideos.com

Edited by Elizabeth Ferretti

All photographs by Peter King unless credited otherwise.

Design and layout
Nikki Moore, Pigsty Studio
www.pigstystudio.co.uk

Printed and bound in Great Britain by Bath Press

Front cover photograph: Traditional Hereford, Westlake Julia, conceived by use of semen from a famous Hereford bull alive in 1964.

Back cover photographs:
Top right: Native Angus.
Top left: George, the English Longhorn.
Middle: Redpoll
Bottom: White Galloway. (Photo courtesy of M. Oertel.)

Frontispiece: Conservation grazing,

Dedication

This book is dedicated to the late Geoffrey Cloke , a master breeder of pedigree livestock, mentor and friend for over 20 years

A Cow Keeper in Golden Lane, Barbican, in the City of London
Illustration by Sir George Scharfe 1862. Illustrated London News, Christmas 1961

Contents

(1) 125, 1st Prize, Scotch Highland Breed. —(2). 99, 144, 72, Sussex Steers, Scotch Polled Heifers or Cows, Shorthorned Heifers.—(3). 329, 310, 291, 297, 220, Fat Wether Sheep of any Speckledfaced Mountain Breed, Fat Wether Sheep of Cheviot Breed, Fat Wether Sheep of Shropshire Breed, Fat Wether Sheep of Oxfordshire Breed, Fat Ewes of Kentish or Romney Marsh Breed ; 360, 364, 355, 350, Pigs of any Black Breed, Berkshire, of any White Breed, White Long Hair Breed.—(4). 41, Hereford.—5. The Anxious Moment. Which is it to be ?—(6). Mr. Sidney showing the Prince of Wales his Prize Taker

THE CATTLE SHOW—LEAVES FROM OUR ARTIST'S SKETCH-BOOK

Taken from the Illustrated London New, December 1877.

iv

Foreword

The end of the twentieth century witnessed a fundamental shift in the emphasis of support for livestock farming in Britain. 'Production, production, production' – the exhortation that drove farmers in the post-War years – was replaced in an accelerating welter of legislation by cross-compliance and the ascendancy of biodiversity and environmental protection. A new range of opportunities has emerged, and native breeds of cattle are poised to take advantage of them.

The author has chosen a wide canvas on which to paint the colourful qualities of native cattle. He catches the mood of the moment, and will infect the reader with the eloquence of his enthusiasm. His free-flowing style runs with breathless exuberance through the attractions of native breeds, moving easily from heritage value to management systems to conservation philosophy.

Cattle originally were domesticated in the Middle East more than ten thousand years ago, and they have been an integral part of the culture and history of the British Isles for more than three thousand years. They have fulfilled many roles, from the use of White Park cattle (Britain's most ancient breed) in the sacrificial rituals of the Druids, to medieval draught oxen, to Britain's fame in the twentieth century as 'the stockyard of the world' based on the export of breeds such as the Angus, Hereford and Beef Shorthorn. Now the emphasis is directed more to their native adaptation for the benefit of the environment. Conservation grazing projects in the Yorkshire Dales prioritise use of local Northern Dairy Shorthorn cattle, while Devons are favoured on the rolling pastures of their native county.

Following reform of the Common Agricultural Policy (CAP) of the EU, and the re-positioning of rural support by Government, livestock farmers are finding it necessary to redefine their objectives and their priorities. Some will take the opportunity to escape entirely from the shackles of unprofitable conventional cattle farming; others will elect to expand their business and manage their animals intensively for even higher yields, and seek to survive through economy of scale; others will understand the new regime and recognise the opportunities for traditional native breeds. The latter group will empathise with the optimism of the author, and others could with benefit review their decisions after they have considered his arguments.

Changing priorities have been accompanied by changing ownership. In recent years the exodus from the land has accelerated, but has been counter-balanced to some extent by an influx of owners without previous experience of farm animals. They bring a new vision to the livestock industry, and a freedom from preconceived ideas – both of which have inestimable potential value – but they lack the basic knowledge of livestock management. They will find that knowledge here. They will be able to plan their systems of production, standards of nutrition, options for marketing, and other aspects of a successful livestock business, from the advice and guidance offered in these pages.

In the last decade, the cattle industry in Britain has suffered two devastating blows. The BSE crisis of the 1990s found its source in the dairy industry, but wreaked its greatest havoc among beef cattle where the majority of native breeds are found. Yet it is interesting that several native breeds escaped almost unscathed from BSE. The arguments rage regarding resistance to disease, but the experience of these breeds may be more than coincidence and deserves closer scrutiny. The Foot and Mouth Disease (FMD) epidemic of 2001 followed closely on the heels of the BSE problem. This time some rare breeds in the eye of the storm were not so lucky. A large proportion of Belted Galloway breeding stock was lost, and it is clear the greatest threats to native breeds in the future are likely to come from disease (and disease control measures) rather than from other causes. BSE and FMD exercise the political process, but many other diseases lie within the control of individual livestock keepers. The author's attention to animal health is timely and appropriate, and will serve to emphasise the importance of biosecurity.

The resurgence of native breeds in the United Kingdom owes part of its success to changes in Government policy. These are welcome changes, and may serve to redress the imbalance of disinterest in earlier decades, but the greatest incentive and support for native breeds has been provided by the Rare Breeds Survival Trust. The Trust has been a strong and consistent champion of native breeds for more than thirty years, working with breed societies and individual breeders in a shared endeavour. On a wider global stage, the Trust also can bask in the reflected success of Rare Breeds International which monitors British breeds abroad. I had the privilege to be Founder Chairman of Rare Breeds International and a founder of the Trust, and share with the author a common interest in these organisations. He worked with me for several years when I was Director of the Trust, and I know the vehemence with which he espouses the opinions expressed in this book. It is a vehemence born of conviction and justified by experience.

As we look to the future, we can be permitted a slight smile of satisfaction. The years are gone when traditional native breeds were dismissed by the mainstream livestock industry as an amusing irrelevance. The breeds described by many as 'museum pieces' now are taking centre stage. They retain their heritage value, but also they are particularly suited to the extensive systems of management which deliver higher standards of animal welfare, greater protection of the environment, and products of higher quality. Their beef is found in the most prestigious restaurants, and sought out by discerning gourmets. As more consumers come to recognise the soft sweetness of beef from the Belted Galloway, or the rich flavour of a White Park 'Sir Loin', the traditional native breeds will flourish. This book is a celebration of their future.

Lawrence Alderson CBE

October 2004

Acknowledgements

Over the years I have been helped and encouraged by some wonderful people. Most of them are still out there doing what they do best, trying to breed and raise the best traditional pedigree cattle they can. Some are being succeeded by the next generation, of which you could be part. I hope you share my good fortune in meeting up with them or their counterparts wherever you are:

All at Archant Publishing, publishers of Country Smallholding Magazine, particularly Editor Diane Cowgill and Karen Foreman (Miss Fixit); my editor for this book, Liz Ferretti, who I owe more to than thanks; the designer, Nikki Moore of Pigsty Studio, for dealing so well with a moving deadline and Ruth Tott of Farming Books and Videos for her vision and unquestioning support for the cause.

The one and only Cloke Family of Chadwick End ,Warwickshire (Traditional Hereford, British Polled Hereford, Dun Galloway, White Galloway and too many others to mention); John & Penny Hawtin, special friends, of Olde House Traditional Herefords, Northamptonshire; the Ladbroke team including Mr & Mrs Rutherford, the Teversons, the Dodds , the Websters, the Winkfields and Brenda and Alan Timms; Stella Scholes of Bovine genetics; Frank Sutton, Parc Gras Longhorns and White Parks Monmouthshire, Wales; Sue Hearn, Trevaskis Gloucester Cattle and British Friesians, Cornwall; Pat & John Holloway (Traditional Herefords and White Park) Wiltshire; Sabine and Joe Zentis, Castle View English Longhorns, Germany; Peter & Jane Symonds, Llandinabo Traditional Herefords, Herefordshire; Paul Dowlman, Head Stockman, Llandinabo Traditional Herefords; the late Major M Symonds, Llandinabo Traditional Herefords; David Powell, Street Traditional Herefords, Herefordshire (the herd has been established over 160 years and he doesn't look a day over 21!); the late Jonathan Blackburn, for having enough faith in me to give me 150 traditional breed cattle to play with and a free hand; Reg Yeomans, a real Devon cattle man who has forgotten more than most of us will ever know; John & Linda McCaig, for their unlimited commitment to Shetland cattle; Lawrence Alderson of Dynevor White Parks and RBST for his support and encouragement through both interesting and difficult times; the Hereford Cattle Society GB and all Hereford breeders everywhere, but particularly members of the Traditional Hereford Breeders Club; the late Mel Railey of El Coyote Texas Longhorns, USA; Don Bixby and Phil Sponenburg of ALBC for all their help on the USA section; Dan Price-Jones, Rare Breeds Canada; Pat Stone of Stoneycreek Texas Longhorns, South Carolina, USA; Heather & Hamish Wilson, Scottish Rare Breeders; David Bradley, Shelly and the rest of the team at Temple Newsham Estate, Leeds for their enormous commitment to conserving rare British cattle in a big way; Peter Ragg; the late MA Thompson; Sophie Grigson the best cook in the world and a great friend; Gordon & Andrew Williams; Bell & Partners Veterinary Group, particularly Ross Butler the best vet I have ever worked with; Dan & Linda Bull; the Whybrow family; Nigel Smith, stockman extraordinaire; Lin (best farm secretary in the world), Sheila Dillon, Producer of Radio Four's Food Programme; Eileen Hayes, Richard Lutwyche, Ann Petch; Catriona and Karen; members and officers of the European Livestock Alliance; RBST support groups, staff, council and members.

Special thanks to Peter and Jane Symonds

Last but not least, my family, especially wife Michele, Caroline and daughters Issy and Emily, and the gang in Cornwall – our love of cattle is truly a family affair.

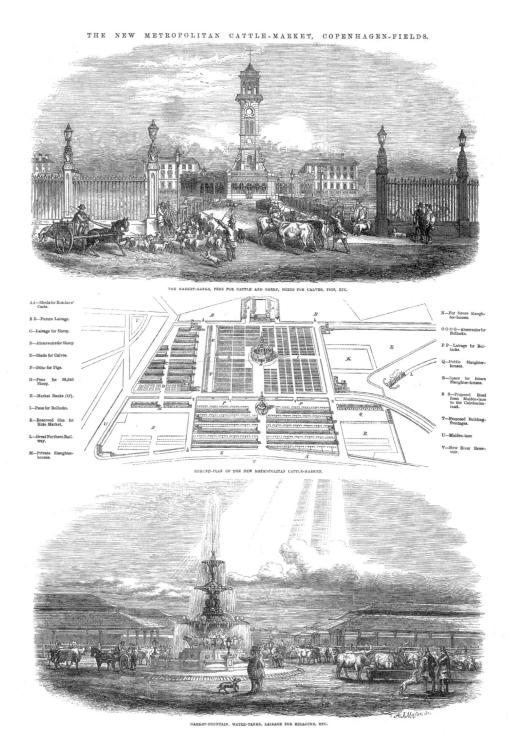

THE NEW METROPOLITAN CATTLE-MARKET, COPENHAGEN-FIELDS.

THE MARKET-BANKS, PENS FOR CATTLE AND SHEEP, SHEDS FOR CALVES, PIGS, ETC.

AA—Sheds for Butchers' Carts.

B B—Future Lairage.

C—Lairage for Sheep.

D—Abreuvoirs for Sheep

E—Sheds for Calves.

F—Ditto for Pigs.

G—Pens for 39,340 Sheep.

H—Market Banks (12).

I—Pens for Bullocks.

K—Reserved Site for Hide Market.

L—Great Northern Railway.

M—Private Slaughterhouses.

N—For future Slaughterhouses.

O O O O—Abreuvoirs for Bullocks.

P P—Lairage for Bullocks.

Q—Public Slaughterhouses.

R—Space for future Slaughterhouses.

S S—Proposed Road from Maiden-lane to the Caledonian-road.

T—Proposed Building-frontages.

U—Maiden-lane

V—New River Reservoir.

GROUND-PLAN OF THE NEW METROPOLITAN CATTLE-MARKET.

MARKET-FOUNTAIN, WATER-TANKS, LAIRAGE FOR BULLOCKS, ETC.

Illustrated London News, June 1855

Introduction

The purpose of this book is to inspire those who have always wanted to keep cattle to get on and do so; to encourage beginners and experienced cattle keepers to favour traditional and rare breeds in whichever part of the world they live; and to encourage people to keep registered pedigree animals, or on larger scale commercial enterprises to keep crossbreeds made up at least in part from traditional or rare breeds.

I have deliberately restricted the breeds I wish to cover in this book as those I am featuring are:

a) proven quality breeds that can be kept with the least trouble;

b) have an increasing value in many cases for pedigree use and set you apart, with very positive consequences, from more mainstream–mass market breeds and

c) because in many cases they deserve far greater support and a higher profile than they ordinarily receive.

Older breeds became established because of their particular merits, be it local adaptation, thriftiness, hardiness, docility, meat quality and so on. Many breeds and types have been altered irreversibly in the drive towards increasing speed and efficiency of production – a laudable aim in manufacturing – however, in agriculture, anybody other than the largest, well-financed commercial producer following this path will be vulnerable to the vagaries of local and global trade, may be under increased pressure to compromise on health and welfare issues, and end up with a product of no more than average quality, designed for the mass market.

In the UK, some producers who have managed to restart their businesses after the catastrophic outbreak of Foot and mouth (FMD) during 2001 have taken the opportunity to launch into new directions. Rather than continue to produce crossbred meat for the less discerning mass market, they have seized the opportunity to reinvent their enterprises. Instead of trying to compete with neighbours whose operations had the advantage of economy of scale, they opted for niche or high quality enterprises using traditional breeds and less intensive management systems, many are beginning to enjoy success, as well as a more satisfying way of making a living off the land.

I have used the word 'traditional' to describe the breeds in this book. Traditional is a broader term than the words 'native' and 'rare' and covers pure–bred animals from breeds with a long, established history and defined or individual characteristics. 'Native' excludes breeds which have established new populations all over the world and 'rare' only covers breeds whose numbers are very low. Generally I do advocate the use, wherever possible, of rare and native breeds but not to the exclusion of all the other sensible possibilities.

Traditional breeds are once again proving their worth. Conservation grazing, for example, places breeds such as the Galloway in areas where more 'modern' breeds would not prosper, yet they thrive, and in doing so conserve important areas of marginal land. The by–product is fine beef and an increasing demand for breeding stock, which in turn helps to secure the future of the breed and the breeders. The Galloway is only one example of traditional breeds of cattle and sheep which are being used in this way.

The UK is in a unique position in the conservation of domestic farm animals because of the diversity of breeds and types which still exist here. In addition, the breeds from our islands have done more than any other to influence domestic animal breeding throughout the world, from the 1900s the UK proudly proclaimed itself 'the stockyard of the world'.

By 1973 the situation had changed dramatically however and a significant number of breeds had quietly disappeared from the world forever. In that year a small group in the UK helped put a stop to this decline by establishing the Rare Breeds Survival Trust (RBST). Since its inception RBST can be proud of the fact that no more breeds have been lost in the UK. The Trust has also inspired the establishment of similar organisations in other parts of the world, including Rare Breeds International, which coordinates global initiatives. Although these organisations have been instrumental in saving a number of rare breeds this would not have been achieved without the support of the people who keep them and breed them. You could and should be one of them and not simply for altruistic reasons, but because, in my experience, traditional breeds can offer the best possibility of creating a successful cattle enterprise.

I have included a certain amount of technical information, but for more detailed study please refer to the superb reference books available on keeping cattle which are listed in the Further Reading section. I aim to provide an overview, which I hope will demystify the subject, and encourage people to keep traditional breeds. This book is largely based on articles written for the UK publication, *Country Smallholding*, and I am grateful to the editor and her assistant for their encouragement in bringing my views to the wider world. I am pleased to say that these articles have resulted in the successful establishment of a number of traditional breed herds here in the UK.

I have also included some details for other organisations who you should contact for more information on breeds and breeders in your area (see Contacts section). If you are outside the UK your national agricultural organisation, should have many contact details. A huge amount of information is also available on the Internet.

Breeding and keeping these traditional breeds of cattle is great fun and incredibly fulfilling.

Peter King,
Herefordshire, May 2004

The author with Frank Sutton's White Park oxen team, Tan and Dawr (meaning fire and water).

A poem on cattle by William Youatt, 1834

She's long in her face,
She's fine in her horn,
She will quickly get fat
Without cake or corn,
She's clean in her jaws
And full in her chine
She's heavy in flank
And wide in her loin.

She's broad in her ribs
And long in her rump
A straight and flat back,
Without ever a hump;
She's wide in her hips
And calm in her eyes,
She's fine in her shoulders –
And thin in her thighs.

She's light in her neck
And small in her tail,
She's wide in her breast
And good at the pail,
She's fine in her bone,
And silky of skin,
She's a grazier's without,
And a butcher's within.

From Cattle, their Breeds, Management and Diseases (Robert Baldwin, London 1834)

3

The White Urus. Illustration by J.Stewart.

Chapter One

The Origins and History of Cattle

It is perhaps surprising that the domestication of cattle preceded the domestication of the horse. Initially, cattle were used more for religious and ritualistic purposes than for their meat, milk, draught or hides. Domestication and adaption of cattle and oxen by man began more than 10,000 years ago in the Middle East; with the process beginning in Europe some 2000 years later. It is interesting to note that some populations in Africa, Eastern Asia and Southern Europe did not utilise the dairy potential of their cattle at all, as a result in some parts of the world people are still unable to absorb lactose in fresh milk.

All existing domestic cattle breeds in the world descend from *Bos primigenius*, also known as Urus (see Plate opposite) or Aurochs (see Figure below). The last European wild Auroch died in Poland as recently as 1627. Originally, the predominant factor in the distribution of the various types of cattle was not so much their suitability to any particular geographical area, but rather the migration of human populations for trade, cultural exchange and territorial expansion.

DNA 'fingerprinting' and other scientific techniques provide a much clearer picture of how today's distinct breeds came into being and, most importantly, how they may be related, as well as measuring the 'genetic distance' between them. Science has helped enormously in unravelling the secrets surrounding the development of cattle breeds but there are still many questions to be answered.

Cattle in Ancient History

- In Egypt, the cow was sacred to the god Isis. Here cattle were gaily decorated, were a highly prized gift and an object of worship.
- Civilisations including Hindu, Greek, Egyptian, Roman and Hebrew used their cattle for parades and pageants and festooned them with wreaths and garlands as well as using them for draught work.

The Auroch

- The Hindus and Hebrews had strict rules protecting cattle from abuse by the owner, such as 'Thou shalt not muzzle the Ox when he treadeth out the corn'.
- Roman writers were particularly insistent on the humane treatment of oxen 'The length of a furrow shall not exceed 120 paces, or else that oxen shall have a time for breath', the ploughman is required to 'shift the yoke, that their backs not be galled' and 'moisten their mouths with water, and to strengthen them with wine when suffering from fatigue' (One can only guess what the response would be if this last instruction was incorporated into the latest animal welfare legislation!) To destroy them wantonly was a punishable crime against the state.
- The Roman word for money was *pecunia*, from *pecus*, cattle.
- Cattle were prized possessions throughout ancient times. The bible states 'if a man shall steal an ox, or a sheep, he shall restore five oxen for an ox and four sheep for a sheep'.
- Hannibal made clever use of oxen to enable his army to escape from the Roman general Fabius. 'Two thousand Oxen, with lighted pine torches tied to their horns, were driven into the hills at night, thus presenting the appearance of a moving army.'
- There is reference to white cattle being used for ritual worship of Celtic deities 2000 years ago. The Celts also used them as a form of currency and to pay fines.

When you consider the above, it is easy to see why cattle have remained such an important part of our history and culture.

Transition – from domestication to selective breeding

At the point of the earliest domestication, cattle groups were not defined as breeds but as types. Early man was still mainly nomadic and had no need to confine these animals, he would follow them using their instincts to help find food. This was also the way with American Indians who followed the herds of buffalo for the same reason. It was only as man developed from being a hunter gatherer and became less nomadic that we started to conserve forage areas to save the cattle having to seek food for themselves. The establishment of settlements and expansion of static populations some 5000 years ago saw the beginning of selective breeding. It became obvious that the animals responded well to being fed generously and could be bred for better conformation and useful production traits.

From type to breed

Man then set about consciously creating breeds of cattle. The Romans were serious students of breeding types; attention was paid to matings, bloodlines and individual animals were identified by branding. After the fall of the Roman Empire the next profound differences in the exploitation of cattle in Britain came several centuries later with the Enclosures Acts whose effects were felt particularly during the 1760s and 1770s. This process of increasing exploitation of cattle continued during the Agricultural and Industrial Revolutions of the eighteenth and nineteenth centuries. The numbers of working oxen began to diminish as horses became more important and then as increasing mechanisation reduced the need for draught animals altogether. People started to move to the towns to look for work and the increasing urban populations demanded more and more meat and milk. These radical events provided the impetus for the 'fixing' of breeds and breed traits and spawned the first generation of 'master breeders' or 'improvers' in the British Isles.

Selective breeding had reached such a level in Britain that in 1856 American writer, Ralph Waldo Emerson was prompted to write;

> *'The native cattle are extinct, but the island is full of artificial breeds. The agriculturist Bakewell created sheep and cows and horses to order, and breeds in which every thing was omitted but what is economical. The cow is sacrificed to her bag; the oxen to his sirloin.'*

It was no accident that Britain was ultimately to be called 'the stockyard of the world' indeed we are still reaping the benefits of the pioneering work of these 'improvers' today. Many of the breeds established during this period, albeit somewhat changed, are still with us – and in some cases this is little short of miraculous.

European cattle types and breeds

European breeds descend from four main groups. The list below includes some of today's most popular breeds together with some rarer examples.

Podolic

Area: lower and mid–Danube, Balkans, Italy
Description: long heads with lyre–shaped horns, large size and 'rangy' (not well fleshed) conformation.
Colour: there are many colour variations
Descendant breeds include: Grey Steppe (Hungary and other countries), Maremmana (Italy)

Grey Steppe. (Photo courtesy of Lawrence Alderson).

'Pattern' group

Area: Germany, Austria, Switzerland, eastern France
Description: short head, slightly concave, medium/short horns curving upwards, medium/large size
Colour: red, black, pied and various colour combinations.
Descendant breeds include: Simmental (Switzerland), Pinzgauer (Austria) Fleckvieh (Germany)

Pinzgauer

Central Europe

Swiss group
Area: Alpine region, northern Italy
Description: short head, medium/short horns curving upwards, medium size
Colour: mainly shades of grey and brown
Descendant breeds include: Brown Swiss (Switzerland) Garfagnina, Pontremolese (Italy)

Brown Swiss Cow. (Photo, Eileen Hayes).

North European

Area: Poland, Czech Rebublic, Slovak Republic, Germany, Denmark, 'Benelux', England

Description: short head, deep forehead, short horns often turned down

Colour: mainly red but also black, brown and pied

Descendant breeds include: Angeln (Germany), Danish Red (Denmark), Friesian (Netherlands), Lincoln Red, Sussex and Shorthorn (England). The Shorthorn type from England had a widespread influence on breed development in France and other countries from the mid to late nineteenth century.

The Sussex. Illustration from Professor Low's Illustrations of the Breeds of Domestic Animals. Courtesy of the Rare Breeds Survival Trust.

'Yellow–Brown' group

Area: central Germany through to France, northern Spain to Portugal

Description: short head, concave face, medium horns but with variation, variable size

Colour: yellow, red, brown

Descendant breeds include: Gelbvieh (Germany), Limousin, Blonde D'Aquitaine, Aubrac, Parthenaise (France), Leonese, Pyrenean (Spain), Mirandesa, Minota (Portugal)

Gelbvieh. (Photo, Eileen Hayes).

Western Europe

Area: SW Iberia, Wales, Scotland, Ireland.

Description: long head, sometimes convex, long spreading horns curving upwards, this breed group is of variable size

Colour: black is very common but also red, pied, and other patterns, especially coloured points. Variety in colour persists in some breeds but selection which took place as many as 2000 years ago can still be seen for example in the coloured points on the ears and muzzle of White Park cattle

Descendant breeds include: White Park, Kerry, Welsh Black, Longhorn, Highland (British Isles) Berrenda, Retinta, Negra Iberica (Spain), Camargue (France)

White Park Ox.

(Much of the above information is taken from Alderson. L., The categorisation of types and breeds of cattle in Europe, Archiovos de Zootecnia, Vol 41, No 154, pp 325 – 334)

Chapter Two
The Birth of the Breeds

Droving from Scotland to London. Illustration Holly Leaves Magazine 1951.

The process of selective breeding of cattle began long before the breeds were 'fixed'. By gradually fixing the breeds man aimed to increase and maintain the animals' desirable characteristics such as docility or good conformation, as well as eliminate any undesirable traits such as poor hooves, a weak constitution, too thick or too thin a hide or a lack of flesh or milk depending on the intended role for the animal.

As far back as neolithic times local or county 'types' appeared which then became the foundation for the later breeds. These 'types' were animals which had been bred over generations in particular geographical areas and might be exceptionally hardy in the hills and mountains and able to withstand extreme wet and cold, such as the type which became the Highland breed. Another example are those types which had the ability to thrive on low quality forage on the plains of eastern England, with poor shelter from the harsh winds from the east coast and became the Lincoln Red and the Red Poll. Although meat animals were moved over great distances, from as far as Scotland to London, breeding cattle were rarely moved far from their home range.

Many of today's pure breeds can be attributed to a relatively small number of cattle breeders or 'livestock improvers' as they were often known, working from the late seventeenth century onwards. We will never know the precise identity of many of these breeders, and in most cases there are no records of the transition from type to the origination of a breed.

The drive for better breeds

Several factors provided the impetus for the development of 'purebred' cattle from the earlier 'mongrel' types. The first was an interest in developing faster racehorses. This started in the sixteenth century during the reign of King Henry VIII when Arabian horses were imported into England for selective breeding. As a result there was better knowledge of the processes of selective breeding and a more organised approach to breeding followed. The second factor was an increase in the size of the human population and a greater demand for food which resulted

Painting of the famous Durham Ox.

in the need for better quality animals who produced meat in larger quantities. The third was a dramatic improvement in farming methods through mechanisation.

More efficient farming methods made it possible to produce winter fodder crops and conserved forage in greater quantities and it therefore became possible to feed greater numbers of animals – winter was no longer a time of starvation. In turn all the extra manure produced by these animals led to increased crop yields.

Country fairs and agricultural societies

During the eighteenth century, local country fairs became the first 'shop windows' for serious breeders in the UK. At first, sheep tended to dominate these gatherings but the importance of cattle steadily increased. Originally, cattle had been selected for draught work with milk and meat being secondary, however a number of events began to change this.

By most accounts, cattle of the 1700s were huge beasts, although they took a long time to fulfil their growth potential. The famous Durham Ox (see Figure above), born in 1796, which toured Britain for six years, eventually weighed 34cwt (3,808 lbs or 1,731 kg). Herefords and Devons were said to weigh over one imperial ton when killed.

As communication improved throughout the UK and news of these prize animals spread, they began to be exhibited farther afield and attracted great publicity. People marvelled at the sight of such imposing animals and stockkeepers began to ask searching questions. How long did it take to reach its weight? What was it fed? At slaughter, in what proportion were meat, fat and offal? These were commercial questions from commercial farmers.

Keeping detailed stock and breeding records became increasingly more important, especially if you were to encourage others to buy stock from you. Breeders needed to establish a good reputation to be regarded as any more than simply great showmen or cattle dealers.

Out of the country fairs were formed the first agricultural societies, who were to provide the ideal venues for 'trail blazing' breeders to exhibit their best animals and so help secure sales and orders. The first show to be organised by one of the agricultural societies was the 'Bath and West of England' in 1777. A report of the show stated that 'several of the nobility and 200 farmers from various parts of the Kingdom were present'.

The first breed descriptions

At the end of the eighteenth century the selection emphasis had begun to change as improved meat and milk production became greater priorities. The quality of early maturity also crept into the breeding equation and record keeping became even more detailed. The following is an early quotation which helps to put these changes into perspective.

> *'In the minute–book of the Kilmarnock Farmers Club there is a report dated 7th August 1795, of a discussion opened by Gilbert Burns (brother of Robert) on "What may further be done to improve the cattle in this country."*

The summary was:

> *"That although much has been done of late in this country in selecting proper individuals of the species to breed from, yet much remains to be done. That particular attention ought to be given to the whole form of the animal as well as to its colour and its horns. That much attention ought to be given to the selection of the cow as well as of the bull. That young cattle, while in a growing state, ought to be more liberally fed than they too generally are in this country; and that as great a proportion of succulent food as possible ought to be given to them in the winter while they are calves, and thereafter plenty of ryegrass hay each spring."*

Another discussion in the same club went an important stage further.

> *"The particular form of cattle the Ayrshire farmer ought to select to breed from"*

Below is the minute which could be said to describe the 'birth' of the Ayrshire breed.

> *"Long and small in the snout, small horns, small neck, clean and light in the chops and dewlap, shortlegged, large in the hind quarters, straight and full in the back, broad above the kidneys and at the knuckle bones, broad and wide in the thigh, but not thick hipped, a thin soft skin of fashionable colours, whatever these be, and the mother carrying her milk pretty high and well forward on the belly."*

These breed descriptions were to become far more detailed and were eventually known as 'points of the breed'. They were the standards by which all pedigree animals would be judged, in the show ring, and perhaps even in the sale ring when farmers were looking for long–term breeding stock and not just a good milker.

Early Cattle Breeders

There were a number of important early cattle breeders whose work over many decades helped to 'make' many of today's breeds.

Robert Bakewell

Robert Bakewell, born in 1725, is probably the best known of the livestock breeders and improvers. His early work was mainly with sheep – the Dishley Leicester, which was originated and perfected by him, eventually made him famous all over the world.

In cattle he is best known for his work on the English Longhorn which he turned into a more profitable animal by improving its beef and milk production. His speciality was 'in and in' breeding. This method sought to fix the best traits of a breed so that they were always transferred across the generations.

Bakewell achieved this through close inbreeding which many at the time regarded at the very least as bad practice and some even thought dangerous. Bakewell did much to change people's perceptions of inbreeding and he went to extraordinary lengths to monitor his results, even to the extent of dissecting cull animals to see how they were made up.

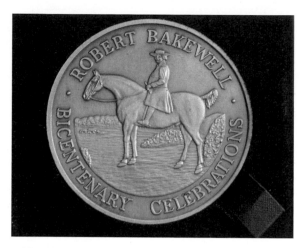

Medal awarded to participants in the Robert Bakewell bicentenary celebrations.

In order to prove his results he pioneered the practice of bull and ram letting, hiring out his animals for breeding. This helped to prove his theories both widely and independently.

Not only was Bakewell interested in improving the productivity of his animals, he was also passionate about temperament. His bulls were well known for their gentle and pleasant disposition – an especially useful trait when dealing with massive animals with long horns. He also abhorred cruel treatment of animals in any way.

Robert Bakewell was an exceptional man and his reputation as one of the most important early breeders of farm animals is well deserved. However, he was far more than simply a livestock improver and it is well worth reading more about his remarkable life and achievements.

The Colling Brothers

Charles and Robert Colling of Ketton Hall, near Darlington, carried out intensive inbreeding with Shorthorns, commencing around 1780. They were so successful that the Shorthorn breed eclipsed Bakewell's improved Longhorn for meat and milk production – a position the breed held throughout the nineteenth century.

William McCombie

William McCombie was born 1805 in the north of Scotland. From 1848 he devoted himself entirely to the improvement of local types developing what was to become the Aberdeen Angus breed. The Aberdeen and the Angus types contained both polled and horned animals, however the differentiation of these two types soon disappeared. During the mid-nineteenth century these black polled animals were threatened with extinction due to extensive crossing with the Shorthorn. (It is believed that the Horned Aberdeenshire breed did become extinct as a result of this crossing). McCombie is recognised for the notable achievement of saving the breed.

There were other breeders, before and after, who played pivotal roles in the development of the Angus, but McCombie is particularly worthy of note because of his fascinating book *Cattle and Cattle Breeders*. If you have ever wondered what it must have been like to drive a herd from

Scotland to London in foul weather, sleep in an inn or by the side of the road with the sale money on your person, along with all the other trials and tribulations of being a successful cattle man in those lawless times, it is all there in his book.

Breed Societies

The first breed societies began to be formed in the mid 1800s. These societies played a vital role in promoting a breed and its breeders to a wider audience at the new agricultural shows, but they also began to attract interest from overseas.

At this time, the global influence Britain had with its empire gave its native breeds a great advantage abroad, and enterprising breeders were quick to capitalise on these opportunities. Stockmen were even sent together with consignments of animals, which had usually been sold for handsome prices, to assist the new owners in setting them up in remote and often hostile environments. The establishment of British breeds in so many parts of the world shows just how effective the breed societies were in promoting them and evidence of their pioneering initiatives is still apparent today.

Herd Registers

Although the shows did much to promote a number of breeds there was a less positive aspect to this, which is still worth noting today. It was all very well seeing a magnificent bull or cow which had been prepared to perfection, but this was not achieved by breeding alone – feeding also played an integral part and quite often these 'show stoppers' had been treated differently from birth. Would the offspring of a prize winner be as good as its parent?

Proving the ability of an animal to transmit its good points to its progeny required data to be built up over generations. This data needed to be recorded in a cohesive way which could withstand scrutiny, as a form of assurance to prospective breeding stock buyers and so the idea of the herd register was born. The first livestock register in Britain was started by Mr Coates for the Shorthorn breed in 1822 and it is still known to this day as *Coates Herd Book*.

Each breed society soon began to publish its own annual herd or stud book. These books contained details of an animal's parentage, prize winnings and eventually included performance data.

George Thomkin's Silver.

Fixing the Breeds

During this time the breeds were a long way from being 'fixed'. Initially 'foundation' animals were registered with the societies and some of their details were recorded. However, it was possible to breed from these foundation animals using other 'types', which may have been of unknown ancestry, and still register the progeny in the herd book. It was not until a herd book became 'closed' to any animal whose sire and dam had not previously been entered that the quest for genetic purity began in real earnest and it became possible to trace an individual animal's bloodlines back to the time when the herd book was closed.

An interesting example of how a breed became fixed is the Hereford. Its red coat and white face are instantly recognisable, but the foundation animals came in many different colour patterns – dark red or brown, roan as well as grey animals with almost no white, while some had mottled faces. The fixing of the white face was a deliberate action of selection which led to the breed we see today. In 1886, at Volume Thirteen, the herd book was closed. The only change to this occurred when it was decided to create the Polled (hornless) Hereford. In the USA, the American Polled Hereford Cattle Club was formed in 1901.

Smithfield champions from the late 1800s, the Illustrated London News.

(Photo courtesy of the North Devon Cattle Breed Society).

Grading Up

An exception to the 'closed' herd book has been a process allowed by some breeds known as 'grading up'. Some breed societies allowed (and some still do) the use of an outside breed for breed improvement, but within strict rules established by the members of the breed society. The process being detailed in the herd book for all to see. Of course illicit crossing has always been practised by less honest breeders but DNA testing has made this less likely, indeed some breed societies require DNA samples as part of their registration procedure.

There is much debate over grading up. In an attempt to compete with other apparently more productive breeds some breed societies have continued to grade up, while others decided to 'close' their herd books decades ago and have maintained the 'purity' of their breeds.

It is easy to understand the pressure on breeds to improve in a highly competitive market, however it should be noted that these commercial pressures have had a tremendous impact on some native British breeds, to the point where in some breeds few completely 'pure' animals exist. The result of extensive grading up is 'improved' animals which are usually larger and more 'productive' but which may have lost some or even all of the characteristics which made them suited to a particular environment. The genetic diversity between the cattle breeds is also significantly reduced.

Of course, a breed cannot stand still, it will wax and wane according to fashion, war, politics and many other influences we could think of. Therefore it is quite an achievement that we still have so many unique breeds in Britain, and even more surprisingly, some have retained their original attributes despite the likelihood of a drift in their characteristics.

Ruby Red Devons are a prime example. It is hard to imagine how such a hardy, thrifty breed which produces great beef could be 'improved' and it is our good fortune that master breeders in Devon have kept them going. It is also pleasing to note that they are not listed as a rare breed.

Maintaining the breeds

To survive in any significant numbers a breed needs two types of breeder – the master breeders and the multipliers. The master breeders found, test and maintain strong bloodlines. This is a highly specialised and increasingly scientific skill, often handed down through the generations of a family together with the cattle themselves. Multipliers use the best cattle produced by the master breeders to increase their stock.

Poor returns in agriculture over many years in the UK, combined with the outbreak of Foot and Mouth disease of 2001, have seen the disappearance of a great number of these 'dynasties'. However, there is a small but increasing generation of new breeders in the UK currently taking advantage of an improving international economic climate – of which you could be one!

Rolla. R. Jones, a Hereford breeder in 1960s America.

Chapter Three

The Breeds

The breeds I have chosen to illustrate here are the best ones to use to establish an enterprise with the potential for a long future, with good economic prospects, for pedigree, meat or dairy potential or a combination of the three.

Some of the traditional British breeds in this chapter are rare and difficult to acquire, but you should investigate your indigenous breeds as a priority in whichever part of the world you are. Contact the breed societies directly for more details and your national agriculture department.

The first part of this chapter deals entirely with British breeds, some of which can be found in other parts of the world, there follows a section on other breeds, particularly from North America. Space does not permit a complete worldwide list, but I have provided contact details for organisations in a number of countries who can help you identify and locate the local breeds which you should consider as a priority (see Contacts section).

Horns

Some of the most visually striking breeds have horns. Beginners may think horned breeds can present real problems, however, from my own experience this is not so. You must always be aware of safety when dealing with any animal and horns should just be seen as part of this equation. Indeed they may even encourage better levels of safety in as much as they are a constant reminder to be vigilant. If you prefer not to keep the horns and you do not intend to show your animals they can be de-horned, but this does dramatically alter the appearance of the animals.

You should not attempt dehorning (also called disbudding) without proper supervision. It is best carried out when the animal is 2–3 weeks old. An accurate injection with an anaesthetic is required so it is a job for the vet until you are fully trained.

Commercial producers have led the drive for polled (hornless) cattle because of the scale of their operations, especially when housing large numbers of cattle together. Polled breeds are an option for the small–scale producer but the choice of breed will be narrower.

Note that horned cattle need more trough and housing space, and any feeders used will need to be the right type to avoid accidents.

Introgression and breed 'improvement'

As discussed in Chapter Two a number of breeds have been used to introduce a variety of characteristics into some of the traditional breeds in order to meet modern requirements and this has resulted in introgression. Occasionally, outside breeds have also been introduced solely for conservation purposes. Most often the latter occurs when the numbers of a breed have become so small that its very existence has been threatened.

Over recent years some of the breeds which were exported from Britain to other parts of the world have been re–imported into the UK as ostensibly 'improved' versions – a well-known example being the Hereford. These breeds were transformed over a very short period, becoming much larger than the original breeds, mainly through the work of skilled Canadian and North American breeders using a method known as 'infusion'.

Infusion is essentially cross–breeding through the blending–in of outside breeds into the original breed, in order to achieve the desired result more quickly than simply by breeding selectively from within the pure population. However, through the infusion process some of the original breed traits may have been lost and with it there is no published, documented record of the outside influences which have been used in the process of improvement – a form of trade secret perhaps? Many people feel however that publicly available breeding records are in fact the best way forward, particularly for breed conservation.

The breeds listed are marked with an (I) if affected by introgression, so that when you approach the breed society and breeders you can specify your requirements clearly.

The most important difference between grading-up and infusion is that in the former case any cross-breeding has been openly recorded in the herd books. The grades (percentage of purity) are adjusted over generations as the degree of purity increases.

Perhaps the most important point for you to bear in mind when deciding between pedigree or 'improved' cattle is that although graded-up animals may be bought for less, their progeny may also sell for less and it might be harder to find buyers. The breed society of your chosen breed can verify pedigree details for you. See Chapter Four for a more detailed discussion on the subject of purebreds, infusion and grading up.

Key to list of breeds

(I) = Introgression (G) = Grading-up Register (R) = Rare Breeds International listed breed.

Beef

Aberdeen Angus (I) Polled

Colour: black (also red)

The Aberdeen Angus, which originated in Scotland, is renowned for beef production. The breed has suffered a high degree of introgression. There is a very small pure nucleus population in the UK and they are both difficult and expensive to acquire, less extreme types are available more cheaply. These may have suffered a certain amount of introgression but still retain some of the characteristics of the traditional breed and may therefore be useful on a smaller enterprise. Early maturing and smaller types are available and are economical to rear for the smaller enterprise. They are very hardy and excellent mothers. Because of the remarkable job the breed societies have done worldwide, the

Traditional type Aberdeen Angus cow.

Aberdeen Angus name is now synonymous with quality and sets a precedent for achieving a premium price. The value of this has been undeserved in some instances as the meat being sold was from a nondescript crossbred whose qualities may not be of the highest standard. Be sure to market yours as pure Aberdeen Angus. Or if crossbred, specify both parents i.e. Angus/Shorthorn, Angus/Hereford etc.

Beef Shorthorn (I) (R) Horned or Polled

This breed also originates in Scotland. Population dipped to fewer than 600 breeding females in the UK, though there were significant populations around the world. In recent times, the breed has enjoyed a surge in popularity in the UK and it is on its way to regaining its well deserved place in the commercial market. Again the breed has been subject to introgression but there are a number of British breeders who specialise in animals free from imported bloodlines, and these are excellent, high quality beef producers.

Beef Shorthorn Cow. (Photo, Eileen Hayes).

This breed can be traced all the way back to the famous Durham Ox mentioned in Chapter Two and the Collings Brothers. Most have a wonderful temperament. They vary in colour from solid red to white, with all the variations in-between, mostly referred to as Roan. A field full of them is a sight never to be forgotten.

British White (G) (R) Polled

This polled breed was classed until recently as a rare breed but its numbers have increased and it is now classed as a minority breed. A grading-up register operated until recently, and there is some debate over which animals are the purest. The British White is also found in Australia and North America.

Originally classed as a dairy/dual-purpose breed, more recently it has been selected for beef production. Those animals which have not been graded up from continental breeds are well suited to range conditions. They all descend from a herd in Lancashire at Middleton Park which originated from cattle acquired from the nearby Whalley

British White Bull (Photo courtesy of the Rare Breeds Survival Trust.)

Abbey. In or around 1765 the entire herd was moved to Norfolk where they became well established. There are still significant herds there, including the Woodbastwick herd which has been in existence for over 100 years. They thrive in hot climates and are resistant to sunburn due to their reflective white coats and dark pigmented skin. They are often confused with the horned White Park breed because of the similarity of their colour markings.

Galloway (I) (R) Polled

Varieties: Belted, Red, Dun, and White

The Galloway is an ancient Scottish breed. It is very hardy, polled and has a thick shaggy coat that together with an undercoat of shorter hair protects it from the elements. Although not a fast maturing breed it produces high quality, well-marbled beef at 30 months or older. You can finish them within 30 months but they may need supplementary feeding. (At the time of writing 30 months is still the upper legal age limit in the UK by which cattle for human consumption must be slaughtered, but in the foreseeable future the limit may be increased.)

Belted Galloway cow and calf. (Photo, Eileen Hayes).

Dun Galloway Cow.

The breed does well on rough grazing and is ideal for hill conditions and areas with high rainfall. It has been used very effectively in swamp areas in Germany and in other conservation grazing initiatives. There are several colour varieties, the Belted, with its white belt, is a listed rare breed in the UK. The White has black points like the White Park and British White but when breeding black calves are produced and black bulls can be used, They are a particularly beautiful breed of cattle.

Young White Galloway bull.

Hereford (I) (R – Traditional type only) Horned and Polled (modern only)

Varieties: Traditional (containing no imported bloodlines), Horned, Modern Horned, Modern Polled

The Hereford is still the most numerous beef breed in the world, however, the vast majority of the animals are of the larger modern type. There are fewer than 450 breeding females of the Traditional type but there are an increasing number of breeders specialising in them. Herefords are excellent converters of grass to high quality beef, are early maturing and renowned for their good temperament.

Traditional Hereford Bull, Llandinabo Usher (Photo,Peter Symonds.)

The Hereford Cattle Society in the UK produces an annual register of the pure nucleus stock – this section of the breed has its own dedicated club. Outside the UK non-extreme traditional types still survive.

Highland Horned

Various colours: dun, reddish brown, cream, brindle, dark brown, black, with wide spread horns.

The Highland is an ancient Celtic breed well adapted to harsh climatic conditions. It is one of the oldest breeds in the UK and more numerous than most people imagine. It produces excellent beef from poor land, but is fairly slow maturing, this needs to be taken into account in the UK while meat animals have to be ready within 30 months because of BSE legislation. As with Galloways, supplementary feeding may be in order, in the UK at least, to get the animals ready within the 30 month limit.

Highland Bull. (Photo, Eileen Hayes).

The dams are known to be very protective of their newborn calves so seek animals of good temperament when starting out. They will respond well to regular contact with their keeper, but if you turn them out on range, and leave them to their own devices, they may be more inclined to urge you to go away rather than come up to you when you enter the field. They are probably one of the most instantly recognisable breeds in the world and will undoubtedly draw a great deal of attention to your enterprise. Some of the best beef I have ever tried came from this breed.

Lincoln Red (I) (R only refers to the pure animals as recognised by the breed society)

Polled and Horned (very few horned animals)

Originally known as the Red Shorthorn, pure Lincoln Reds are in very short supply at the time of writing but some might be available. These are large docile animals able to produce beef from modest inputs. The breed has been successfully exported to many parts of the world, proving its quality and adaptability. They descend originally from the Shorthorn

Lincoln Red

and would lend themselves very well to the modern niche enterprise – they are a pleasure to keep. One bull I know of is inseparable from a Suffolk Punch horse which he is kept with when not in use for breeding. (The only problem is that he does not care too much that the horse is not a cow!)

English Longhorn (R) Horned

The English Longhorn is a conservation success story – it is now classed as a minority rather than a rare breed. Their striking appearance has no doubt helped their increase in numbers in recent years. They are generally large, docile, hardy cattle and produce excellent beef. Their horns come in many shapes and there is a wide range of colour patterns. There is an excellent book in the Further Reading list devoted entirely the breed and its history. (No relation to the Texas Longhorn, see section following on North American breeds.)

English Longhorns in Germany.

One of the best herds of English Longhorn I have seen is in Germany. A friend in Herefordshire runs a very large herd successfully with minimal supervision. It is another British breed that makes you wonder why we had to go to Europe for large cattle. Some Longhorns have extraordinarily large frames, but this type may need supplementary feeding. The good news is that because there are so many good herds you have a wide choice of places to acquire stock. Aesthetically, they are stunning.

Luing (I) Horned

Various colours from red to roan.

This breed was created in the 1940s by the famous Cadzow brothers on the island of Luing off the west coast of Scotland, by crossing Beef Shorthorn bulls with Highland cows. They show good growth rates on poor forage for beef production. There are a few, well established herds in the UK.

Luing. (Photo courtesy of the Luing Cattle Breed Society.)

North Devon (I) (R) Horned and Polled

Often known as Ruby Reds because of their thick red coat, North Devons are very hardy and early maturing. The breed has been very popular in hotter climates around the world where it has been used successfully to create specialised breeds. They produce high quality beef. They are a breed with very ancient origins, well known for their placid temperament and ability to do well on little in relatively harsh conditions.

North Devon Bull Werrington Booty. (Photo, North Devon Cattle Breed Society).

South Devon (I) Horned and Polled

Medium red to yellow red, originally a dual-purpose breed but now a beef breed in the UK. They are large animals also used for draught purposes. High growth rates can be achieved. Some animals have the double muscling potential of breeds such as the Belgian Blue but I would advise the smaller keeper to choose the conventional type. The South Devon is probably the largest of the British breeds, and is ideal if you plan to set up a larger

South Devon. (Photo, Anthony Mosley).
Courtesy of South Devon Herd Book Society

scale enterprise producing a lot of meat. It can and should be chosen above imported commercial breeds. South Devon bulls are used commercially in some herds for cross breeding.

Sussex (I) (R) Horned

This docile, dark red breed has proved popular in many parts of the world. Of a similar size to the North Devons they are early maturing beef producers from lower quality forage. Historically a popular breed for draught purposes. There are initiatives to increase their competitiveness against the large imported breeds. This type of 'improved' Sussex should not be the type for the specialist enterprise. You may have to look

Sussex Heifer. (Photo, Eileen Hayes).

hard to acquire the originals, but it is worth the effort. I have personal experience of this breed, on very poor land in a high rainfall area where they exceeded our expectations.

Welsh Black Horned

These famous black animals originate from ancient Celtic cattle known for centuries. They are hardy animals which were originally a dual-purpose type but they have been selected for beef for a number of years. The cows make excellent mothers, and the animals show good growth rates for beef production. They have enjoyed the loyalty of Welsh breeders over generations, and not purely

Welsh Black Bull. (Photo, Eileen Hayes).

for sentimental reasons. They have had to pay their way, and continue to do so even today, against the fierce challenges of fashion. The breed society has worked hard to support the breed through a successful and consistent marketing campaign. Temperament can be an issue, so bear this in mind during the acquisition process.

Also worth mentioning is an obscure type which is descended from this breed called the Ancient Cattle of Wales or *Gwartheg Hynafol Cymru*. This type comes in a variety of colours, contact the society listed for more details.

White Park (R) Horned

The oldest recorded breed in Britain, first noted two thousand years ago. White Park cattle are striking white animals with black, or less commonly red, points to the ears and nose. They show strong hybrid vigour when crossed (i.e. the offspring's performance is better than that of the parent breeds), due to their genetic remoteness from other breeds.

Fig. 3. White Park oxen

This breed is adapted to both rich pasture and low input systems and is extremely hardy. Excellent producers of quality beef from low quality forage, some say it has a gamier flavour than some other breeds.

One of the most impressive and successful herds I know, lives out all year round on Salisbury Plain, with minimal management. Originally they were prized as animals to be hunted on horseback in enclosed parklands. The great estates often chose them for their stunning appearance. These days the breed has been developed for quality beef production. Beef animals may need supplementary feeding to be finished by the 30-month limit. They can be very large (see Figure on Page 2 of two White Park oxen, Tan and Dawr (Fire and Water), owned by Frank Sutton. It shows the full potential size of the breed. The oxen are 12 years old .)

Dual purpose

Dexter (R) Horned

Colour: black, red or dun:

The Dexter is probably the best example of successful conservation, numbers have increased significantly over recent years. The beef enjoys a strong reputation for quality. Dexters are considered by many to be the ideal smallholders cow, partly because of their small size and thriftiness. They are also very popular as 'house cows' for milk.

The Langley herd of Dexter Cattle. Photo: Stuart Creasey.

The breed has an interesting genetic trait that you need to be aware of – a fatal (or semi-fatal) gene within the breed can result in 'bulldog' calves which are not viable and are usually aborted in the later stages of pregnancy. The breeders and the Dexter society have recently carried out in-depth studies and the phenomenon is now well understood and therefore the risk of it occurring can be minimised. It is most likely to occur when mating short-legged animals, it is therefore advisable that you include the longer legged type in your herd. The Dexter produces smaller joints than other breeds, which can be a marketing advantage when selling directly to consumers. One farmer I know, following the outbreak of foot-and-mouth disease in the UK, recently sold his crossbred suckler herd and replaced them with Dexters, much to the initial bewilderment of his neighbours, however, only two years in, his new enterprise is highly successful and he cannot produce beef quickly enough for the Traditional Breeds Meat Marketing Scheme.

Dexters are fairly plentiful so you should have a wide choice of animals, and they need not be too expensive to start up with, partly due to their size, but obviously the meat yield is commensurate.

Gloucester (G) (R) Horned

Colour: dark mahogany to light red with a white back stripe, tail and underside.

The Gloucester is undoubtedly one of Britain's oldest dairy breeds, made famous from the use of its milk to produce Double Gloucester cheese. Recently the breed has been selected more for beef, however at the time of writing one UK breeder is once again producing cheese from the milk. Much of the breed's milking ability has been lost so you will need to select your animals carefully if this is your main requirement. Gloucesters make excellent 'house cows' and the meat is

Gloucester cow.

beginning to enjoy a reputation for very high quality. If you have any intention of showing, make this clear to the seller as there remains the potential to produce mismarked animals, particularly from those animals which have been graded up through the use of other breeds, the herd owners and breed society can help you with this.

Irish Moiled (G) (R) Polled

Variety of colour patterns based on red through to white.

The Irish Moiled (moiled means hornless) is one of the rarest breeds in the United Kingdom. The majority of today's animals are of the beef type, but some dairy types still exist if you want milk. They make an ideal hardy house cow and do very well at shows partly because of their striking colour patterns. This adds to the commercial potential, some examples change hands for as much as £3,000. The breed was in such low numbers that recently a grading up programme using other

Irish Moiled heifer. (Photo courtesy of Rare Breeds Survival Trust.)

breeds has taken place. Such graded animals should cost less to buy, and represent a cost effective way in to a breed which is low in numbers. There is a breed society and a UK mainland support group who you can contact for further information.

Red Poll (R) Polled

Colour: dark red

A hardy breed. There are only a small number of milking herds left in the UK today due in part to their relatively limited milk production in comparison to the dominant Holstein, however a few dairy herds have survived and one in particular provides the milk for Welsh Llanboidy cheese. In an endeavour to increase yields some breeders outside the society have begun using other breeds in their selection programmes, however, you should find pure milky animals through the society, as this diminishing section of the breed really needs support. The Red

Red Poll bull. (Photo, Simon R. F. Tupper. Courtesy of the Rare Breeds Survival Trust).

Poll is a very useful all-round animal – they are a good breed for multiple suckling if you want to rear extra calves cost effectively and are excellent mothers, they also produce high quality, fine grained beef economically and are early maturing. They are the living descendants of the now extinct Suffolk Dun and Norfolk Red. They can do a good job of keeping you, rather than you keeping them.

Shetland (R) Horned

Colour: black and white, more rarely red and white

This small, hardy 'crofter's cow' is ideal for the smallholder. They thrive on poor quality grazing. They tend to reach a greater size on the mainland than they would on their native isle due to the less harsh environment and longer access to good grass. The Shetland is useful for multiple suckling purposes and produces good beef. As well as having their own society in Shetland, a very active mainland breeders group exists for support and advice. The largest herd is not on the islands but on the Scottish mainland.

Shetland cow. (Photo, Rare Breeds Survival Trust).

One positive trait, which nearly made them extinct, is their potential for commercial crossbreeding. The result was that too few pure Shetlands were being bred. If you want lots of meat, you could establish a pure nucleus herd to ensure a supply of replacements, but crossbreed from a proportion of the herd using, for example Shorthorn or Hereford for faster maturing larger carcases. (Continentals such as Simmental are also used very successfully, but I am advocating British crosses).

Dairy

Though most newcomers to cattle keeping will be more interested in beef breeds with perhaps some milk, some will be interested in producing cheese and other dairy products. The following is a list of dairy breeds to consider. In any number the larger breeds such as the Ayrshire are for larger enterprises, but an individual one would be suitable as a house cow if you have enough uses for the copious amounts of milk they can yield.

Ayrshire (I) Horned

Colour: brown and white.

The Ayrshire was originally well known for producing good quality milk from low quality forage. Unfortunately the breed has suffered from severe introgression from the modern Holstein and so it may be quite difficult to identify traditional types. I list them here for those thinking of establishing a larger dairy-based enterprise. They were plentiful in the 50s and 60s, but are much less common today. They are definitely worthy of your consideration and need support to maintain their purity from crossbreeding.

Dehorned Ayrshire cow. (Photo, Eileen Hayes).

British Friesian Polled

Colour: black and white

Although strictly speaking an imported breed, in my opinion the British Friesian represents an important genetic resource in the UK, not least because of its link to the Hereford for producing what were colloquially known as 'Black Herefords'.

The breed was the mainstay of the British dairy industry between circa 1950–1980 before the modern Holstein took over. A significant number of breeders have remained loyal to the British Friesian and there is a club

Friesian cow. (Photo, Eileen Hayes).

dedicated to their continuation and promotion. They may also be classified as a dual-purpose breed in that they have good conformation and so produce good beef calves when crossed with a bull such as a Traditional Hereford – this has been the most common combination. Cull cows also have a meat value at the end of their long working lives, although this will not be relevant again until meat from over 30 month old animals is allowed to enter the food chain again.

Channel Island Breeds

Jersey (R) (I) Horned

Colour: various shades of fawn

Jersey cow. (Photo, Eileen Hayes).

Jersey milk, sold in the UK for many years as Gold Top milk, is still available in its homogenised form and is sold as 'Breakfast Milk' at UK supermarkets. Its high butterfat content gives its milk a distinctive creamy quality, it is also prized in the making of rich yellow Channel Islands butter, which is quite hard to find. Jerseys make ideal house cows and are relatively small.

The bulls have a reputation for being quite dangerous, as do most dairy bulls, but most people are unlikely to keep one. Surplus bull calves have traditionally been regarded as worthless and they can be bought very cheaply. These can be reared for meat with reasonably good results.

The Jersey island breed society has protected the breed from introgression by refusing to allow the importation of modern North American genetics. Skimmed and semi-skimmed milk have gained supremacy over recent years, but with more doctors pronouncing that fat is not necessarily bad for you, if you have a balanced diet and exercise, there is room for hope that Channel Islands whole milk products can regain their rightful position in a natural and varied diet! Channel Islands whole milk, or ice cream or butter made from the milk is in my opinion unforgettable.

Guernsey (I) Horned

Colour: golden brown and white

Guernsey cow with calf. (Photo, Eileen Hayes).

Unlike the Jersey, the Guernsey has undergone wholesale 'infusion' in Canada, if you are looking for the original types seek advice from the relevant society. The larger animals will not be the best choice for your unique enterprise. You might get more milk, but will it be as good? How much more will it have cost to produce? Will the cows live as long? Will they have any other weaknesses such as higher mastitis levels? Why take the risk for short term gain.

Dairy Shorthorn (I) (R) Horned (pure nucleus animals only)

This famous breed enjoyed dual-purpose status for the same reasons as the Friesian. They make excellent house cows, are hardy and will do well on poorer grazing. Again this breed has been heavily influenced by North

Northern Dairy Shorthorn cow.

America-style infusion which has resulted in the production of a modern Holstein type. Note that many Dairy Shorthorns are hybrids known as 'Blended Red and White', if looking for the original type seek advice from the breed society. The originals do exist but need support. Dedicated breeders will be very eager to assist you in your search for the right animals. The breed society also has a computerised database which makes identifying the purest animals easier

.Mention must be made of a very special group of animals recognised by the RBST. The Northern Dairy Shorthorn. The remaining few animals descend from a type that were developed in the north of England. A handful of knowledgeable, dedicated breeders are working with this breed. Whitebred Shorthorn As the name suggests this is a white variety produced mainly to provide White Shorthorn bulls for crossing purposes. They have their own Association (see Contacts section).

Kerry (R) Horned

Colour: black

This ancient breed has Irish Celtic origins. These small, hardy, thrifty cattle are another ideal choice for the small-scale producer or as a house cow as they are relatively productive for their size. There are still some milking herds in Ireland and a small number of specialised herds in the UK. Registrations are administered by the Royal Dublin Society, and there is a UK

Kerry cow. (Photo, Eileen Hayes).

Society for breeders which was established in 1882. The Kerry is believed to be the breed from which the Dexter descended. Compared to the Dexter the Kerry is exceedingly rare, but can be acquired for a very reasonable price. The breed is in urgent need of greater support and the fact that they are so rare is a good selling point. If you are on a limited acreage they are a far better option than keeping a smaller number of larger cattle.

Traditional Cattle in North America

I would like to have included a chapter on every country in the world, but that would be a book in itself. I have chosen North America as the best example outside the UK in the hope it will inspire those of you in other countries to investigate your own rich history before going out and acquiring plain old commercial cattle. As in the US, there may be valuable vestigial populations of British breeds imported long ago that you could call your own, as well as fascinating native breeds which deserve your attention.

North America, including Canada, has acted as a time capsule for cattle breeds. It is only little over 500 years since colonists from all over the world began to settle there, and their cattle breeds migrated with them. In the south were Spanish cattle, further north British breeds began to gain a foothold from very early on. The survival of the fittest and most adaptable soon sorted the 'wheat from the chaff'. Those that have survived to today have their own American heritage whatever their origins. Some of the breeds are household names, others are like well-kept secrets, and just as in the UK some breeds have been in danger of disappearing.

The US organisation, the American Livestock Breeds Conservancy (ALBC) was established in 1977 to monitor and support the most endangered domestic farm livestock within the United States. Their listing system and some of their programmes were inspired by the Rare Breeds Survival Trust (RBST) in the UK, but ALBC has developed its own unique strategies and systems to cope with a geographically challenging task. In such a huge continent ALBC's membership is comparatively small for the size of the territory and resources have always been stretched, but their achievements have been remarkable with cattle and other species, particularly poultry. (Details of how to contact or join the organisation are in the appendix, see Contents section.) Canada is covered by its own organisation, Rare Breeds Canada (RBC) and there is a high level of cooperation between them and the ALBC.

In this chapter we shall look at listed rare American breeds. The most famous US breed, the Texas Longhorn, is also included – it is by no means rare, but it is most certainly a traditional breed and it does have a related nucleus population which is very rare.

Florida Cracker

As the name suggests, this breed has developed in Florida. It is descended from the Spanish 'Criollo' cattle which originally went to North and South America. Criollo means cattle of Spanish origin but bred in the 'New World'. In appearance Florida Crackers are like small Texas Longhorns, the mature weight of a cow is around 400 kilos (900 lbs). Their horn lengths are

Florida Cracker. (Photo, D.P. Sponberg.
Courtesy of the American Livestock Breeds Conservancy.)

not as extreme as their Texas cousins, but they do share similar colour patterns. Heifers are known to reach sexual maturity at an early age and their fertility is high. They were specially adapted to Florida scrubland and heavily wooded lowland areas – they could withstand the insects, heat and extremely high humidity. Despite the onslaught of imported breeds, they were still in reasonable numbers until the mid 1950s, but then they nearly disappeared altogether through being crossed with Brahman, Hereford and Angus. Fortunately several families of breeders, in Florida and elsewhere, maintained pure herds because they valued the traits of the purebreds as well as their importance from a heritage point of view.

Since 1970 the State of Florida has also been involved in the conservation of the breed. In 1989, the Florida Cracker Cattle Breeders Association was established. Four hundred animals were used to form the foundation register. Currently they are in the ALBC critical category and are very worthy of your support. If you want beef from a low-input system on marginal ground and some nice hides and horns to boot, the Cracker could be ideal.

Pineywoods

Pineywoods are another breed which owe their origins to the Spanish cattle imported to America from the 1500s which became known as Criollo cattle. The Pineywood was developed to be productive on low-input, highly extensive systems. The stronghold of the Pineywoods was Alabama, Mississipi and Georgia plus other parts of the south-eastern United States. Typically, they were tolerant to heat, were known for their longevity and ability to resist parasites and disease.

This breed was a triple-purpose breed used for beef, occasionally milk, and extensively as oxen for timber hauling. They became rare at the end of the 1800s, displaced by other mainly British and European imports. They were also subjected to crossing which further threatened the pure population and they very nearly became extinct. Once again a few families remained true to the breed giving their name to the strains they developed. The Carter strain, for example, began in the 1860s when Print Carter, a sixteen-year-old civil war veteran, swam a herd of these red cows across the Pearl river and began breeding them. The last animal to enter the herd from outside was a bull in 1895; the herd still exists and is thriving.

Pineywoods are small cattle, with the rugged qualities they need to survive on their home ranges, though their size can vary according to how harsh or otherwise their location is. Environmental factors also govern the breeding selection process. Particularly harsh seasons would influence which animals born the previous year would survive, and which of the adults would

Pineywood. Photo D.P. Sponberg.
Courtesy of the American Livestock Breeds Conservancy

breed successfully. Cows weigh 270 – 360 kilos (600 – 800 lbs), bulls 360–540 kilos (800 to 1200 lbs). Some 'dwarf' cattle are found within the breed. Most are horned, the colour pattern is as wide as that in other animals of Spanish origin. Some specific colours have been selected for, including solid colours such as dun. They are listed as 'critical;' and are in desperate need of support. The overwhelming case for their heritage value can be in no doubt, but why not keep some for your own benefit, and by doing so you will also help to ensure they survive.

This breed descends from a type of lineback-patterned cattle with a distinctive line running along their back which is usually white. These were common in New England. The exact origins of this breed are unclear, although a breed association existed for a brief period in the early 1900s. The original linebacks were triple-purpose, used for milk, meat and as draught animals. There was no formally recognised breed as such, and no breed society apart from a short-lived organisation formed early in the 1900s. Most of them disappeared through being crossed with the more dominant Holstein.

Randall Lineback

The name Randall comes from the Randall family in Vermont – their herd was one of the few which had not been crossed, but unfortunately, on the death of Everett Randall, the herd was dispersed and most were slaughtered. Members of the ALBC were alerted and mounted an urgent rescue plan. A small part of the herd was saved, and it is from these few animals that the current population descends – there are currently fewer than two hundred Randall Linebacks alive.

Randall Linebacks. (Photo,D.P. Sponberg.
Courtesy of the American Livestock Breeds Conservancy.)

Randall Linebacks are medium sized, the cows are predominantly of a dairy type,the bulls are large and have good growth rates. Surprisingly, within such a small population there is significant variation in size and shape. The 'lineback' name refers to the characteristic white line down their backs, they are generally blue-black with a striking roan pattern with all the variants in between. The males make good and willing oxen. The striking colour patterns are definitely a plus point, from an aesthetic point of view the breed is stunning.

It should be noted that Randalls are no relation to American Linebacks (although some of the latter are historic animals, their registry does not represent a genetically pure breed).

While the critically rare Randall is being closely monitored and managed it needs a lot of help, and urgently.

Texas Longhorn

No consideration of American cattle breeds would be complete without including the Texas Longhorn. It is not a rare breed but it is as traditional to the US as you can get. The Texas Longhorn is of Spanish descent and is no relation to the English Longhorn. The Longhorn is best described as 'a breed fashioned entirely by nature in North America'.

The breed thrived in mostly independent herds, on huge ranges, with little support from men and this meant that its self-sufficiency and independence continued to evolve over

Original type Texas Longhorn. (Photo, D.P. Sponberg Courtesy of the American Livestock Breeds Conservancy.)

generations to produce one of the hardiest breeds in the world. One of its most useful characteristics was that it could travel great distances to find food and water. Stories abound of how protective they are of each other and their calves. An inspiring book by Frank Dobie about Longhorns (*Longhorns*, Nicholson & Watson, London, 1943) will leave you with the deepest admiration for the animals and the men who made their living from them.

Longhorns were used to fill the void left with the disappearance of buffalo after the civil war. Herds were brought from the harsh south to the lusher north, they adapted successfully and were used to produce the meat so desperately needed at the time, and at the lowest cost. They were the American breed supreme.

Then imported breeds gradually appeared in the 1800s which were faster maturing and had better conformation, this fact coupled with the fencing-in of the vast ranges which had been the home of the Longhorn, led to a sharp decline in their numbers. This might have been the end of the story but in 1927 the Texas Longhorn was recognised for its uniqueness and a conservation programme was initiated by the US Government on wildlife refuges in Oklahoma and Nebraska. Once again a few stalwarts of the breed who recognised their genetic value had maintained small herds of the original breed. The fortunes of the breed were also improved by the formation of the Texas Longhorn Cattle Breeders Association in 1964.

The breed was originally saved for heritage reasons, however it is satisfying to report that the Longhorn is now being looked at afresh and is being recognised for its commercial potential. The Longhorn represents a unique 'genetic package'. One breeder in the US, for example, has found a market for almost every part of the animal – orders for the meat exceed supply, people increasingly prefer Longhorn reared on range to feedlot beef, but she also sells cured skins, hair on, as well as the horns and skulls. What better example of the need for a longer-term view of the value of domestic farm animals?

In the US the breed has gained so much in popularity that bulky 'show types', very different from the originals, are increasingly being bred. If you have the space try and find some of the older type, you are sure to learn a lot from them, and they won't need much from you!

Canadienne

Canadienne cattle (see illustration, page 62) originated from imports to Canada from Normandy and Brittany during the sixteenth and seventeenth centuries. Later imports also came to the US from Brittany and Gascony. They were bred from and selected for both hardiness and productivity. The resultant type was kept largely pure and became the Canadienne. By the mid 1800s the attributes of the breed became less popular and they were being crossed with larger imported breeds less capable of coping with the natural environment.

In 1895 a group of concerned individuals, including some academics, formed the Canadienne Cattle Breeders Association. In recent times the Quebec government has also lent its support and most of the breeders and cattle are still to be found in this area. Rare Breeds Canada works tirelessly to ensure their survival. The breed does well on little and does not require intensive management. It is predominantly a dairy animal, and makes a superb house cow. A small herd is ideal for making specialist dairy products. They also produce lean meat, and have a very high meat-to-bone ratio, which can greatly help with the economics of keeping them. When born they are pale, they become black or dark brown on reaching maturity.

There are a number of breeds of British and European origin in the USA. Some have been there so long they can be defined as distinct populations.

Milking Devons

It is many years since Devons were milked in the UK, but in the US the 'Milking' Devon has survived. The story is that in 1623 two heifers and a bull were sent from North Devon in Britain to a member of the Plymouth Colony – reputedly the first cattle to be imported to the American continent from Britain. Their main use was as draught animals, a role in which the Devon excelled. In due course more of

Milking Devon. (Photo, Robert Mitchell. Courtesy of the American Livestock Breeds Conservancy.)

them were imported and they began to be used for milk, cheese and butter and ultimately for prime beef production. They are red in colour, ranging from deep to light red or chestnut, very docile, aesthetically stunning and of medium size. As the need for draught animals diminished so specialised dairy breeds began to dominate.

In 1952 the Devon breeders in the US decided that the breed needed to specialise as a beef animal to survive. At the same time another group of Devon breeders decided to commit themselves to keeping the triple-purpose type going. Regrettably that association slowly disappeared, but in 1978 a new registry was founded for the remaining animals which had retained their triple-purpose traits as well as the ability to thrive in harsh environments with minimal management. The efforts of these groups have paid off. Recently, however, two important herds were slaughtered without warning and renewed efforts are urgently needed to regain lost ground.

The breed is now listed by ALBC as 'critical'.

This wonderful triple-purpose breed, producing milk, meat and used for draught if needed, is a truly great breed for the pastoralist, smallholder or top quality niche meat producer and is an important part of the heritage of the founding fathers.

Beef Devon

The Beef Devon has the same origins as the 'milking' Devon but has been selected for beef. It is not to be confused with the larger, yellower South Devon.

The Beef Devon has enjoyed something of an upsurge in popularity of late and finds itself in the 'recovering' category as far as its conservation status is concerned. It is nevertheless highly worthy of your attention if you want to produce top-quality beef.

Other breeds

Other breeds of European origin to be found in the USA are the Kerry from Ireland, which is the small black dairy breed from which the Dexter is descended, White Park, Red Poll, Galloway, Guernsey, Highland, Belted Galloway, Ayrshire, Dutch Belted (Lakenvelder). These are all listed with ALBC. There are of course more numerous breeds such as the British White, Aberdeen Angus, Hereford, Sussex, Shorthorn. None of these in their modern form are under threat, but the original types have long been superseded by the larger modern versions, some of which have suffered introgression from other breeds. In South America, particularly Uruguay, the old type of Hereford can still be found. If you decide to keep one of the more numerous breeds, I strongly urge you to go for the original, more compact, type. They are more likely to keep you than you them.

Rest of the world

Wherever you find yourself there will be breeds which are particularly adapted to your conditions. The household name breeds are present in the farthest corners of the world, but even in places like Brazil, where the Angus and Hereford have so long been popular there are specialised native breeds which warrant attention. In Holland the Lakenvelder should be on your list, in Germany the Glanvieh, in Spain the Cardena Andaluzia is left in only four herds, In Italy the dairy breed Cabannina ..., the list is seemingly endless, but much more is known today about the true position of the breeds than ever before.

In conclusion

It is worth investigating a wide range of breeds before basing your cattle enterprise on the easiest to find and most fashionable modern breeds. By differentiating your business from the big mainstream producers, you will achieve greater publicity, probably produce superior products if you do it well, and you will be doing something to maintain the genetic diversity and heritage of your native breeds. You will also be likely to find that there are other people like you who you can collaborate with.

If you have a national rare breeds or cattle organisation, join it. Where specialist organisations do not yet exist, contact your government ministry of agriculture for breed lists and up to date contact details, animal genetic resources (AnGr) are being catalogued worldwide so it has never been easier to investigate.

Chapter Four

Establishing your herd

Traditional Hereford family group.

For the smallholder or novice who has kept a few sheep, poultry, goats etc., trying your hand at cattle can seem like a huge step to take, it needn't be. I hope that with this book I can help to demystify the job and encourage you to take the plunge. As much as anything, the cattle breeds need you. Be assured, if I had found keeping and breeding cattle overwhelmingly difficult, I would not be prepared to put pen to paper.

My approach is that you know nothing about cattle, if you already have some experience, there will still be items to interest you. If you are an established cattle farmer, I urge you to put aside any rigid views you may have and take a fresh look at the older breeds in a rapidly changing world. They may be able to keep you rather than you keeping them.

I have had the great fortune to know and meet some of the greatest cattle people, they have helped me no end. If I can share some of their knowledge with you, together with my own experiences, it should make your life considerably easier in establishing your own enterprise, whether it be a solitary house cow, a small suckler herd or a larger project.

Almost *because* of the times we are in, I think that it is even more important to urge you to take the plunge. World trade and modern food production methods are placing increasing pressure on standards and systems. Demands from increasing numbers of discerning consumers are beginning to favour those of us who farm with traditional breeds and methods. Those who farm traditional breeds can set a benchmark for high quality in both animal welfare and food quality, based on breeds adapted over long periods of time for that very purpose. In terms of cattle generations, one breeder's lifetime influence is quite small, so new new breeders are needed, worldwide.

Taking the above into account, I want you to favour breeds that are well suited to your environment, and non-intensive management, even organic, if you can meet the standards.

Choosing the right breed for you

Several factors need to be taken into consideration when looking at the breeds available today. The following short list will enable you to do further research and make a fully informed decision.

- The size of your holding as this will have a bearing on the size of the animals you wish to keep, for example you could keep three Dexters to one Longhorn.

- The quality of your land. Is it rich, well-drained lowland pasture with plentiful rich grass, or swampy ground with coarse low nutrient grasses, or even thin soil on an exposed hillside? It is a good idea to gain local knowledge from your neighbours on this, but it can still pay to keep an open mind.

- Are you planning to produce large quantities of meat for manufacturing added value products or prime beef, or are you going to produce meat and milk for you and your family and friends?

Little and Large: Dexter and Devon bulls. (Photo, Eileen Hayes).

- Will you be pursuing the opportunity to sell registered pedigree breeding stock? Some breeds are in greater demand than others.

Basic Requirements

Registering your herd

In the UK, if you acquire cattle you must inform your local Animal Health Department and your local DEFRA office (Department for the Environment Food and Rural Affairs) that you are establishing a herd (even if you have only one cow). You will be issued with a unique herd number. If you already keep sheep or goats you may already have a number but you still need to notify the relevant authorities. DEFRA and your local Animal Health Department should also provide you with guidelines on current legislation and welfare codes for cattle. In other countries check with your local livestock authorities.

Site Visit

In the UK you are likely to receive a visit from a DEFRA official to establish that you have suitable facilities for keeping cattle.

Feed Store

You will need an adequate fodder store for straw, hay / silage. The size of your store will depend on whether you are producing your own fodder or buying in batches as needed. You will need a vermin-proof store for hard feed (corn etc.), such as old freezers.

Fencing

You need to ensure that your fencing is suitable. It does not have to be brand new, but it does need to be good enough to keep your stock in and your neighbours' stock out. Check that there are no parts which could cause injury to you or your animals such as loose strands of barbed wire. Block any gaps in hedges to prevent curious calves escaping, if the cow goes looking for the calf you will suddenly have a very large gap!

Block any gaps in walls to prevent curious animals escaping.

Housing

To be approved by DEFRA as having a holding suitable for cattle you MUST have an isolation facility, you may need to confine an animal in a warm, dry place or for calving, as much for the stockman's sake as the animal, a shelter is better than a torchlight operation in slanting rain at two in the morning. You may also need to provide some wet weather shelter, particularly in the winter, but this will depend on the breed and your type of land. You may need to bring them in for part of the winter if you don't want to damage your pasture

Know your land

If you have not already done so, you should have your soil tested to have a clear picture of the true properties of your land including mineral deficiencies, PH levels etc it will save you a lot of time and avoid you making wrong decisions. It may also be the case that your soil contains too much of some elements, such as iron or molybdenum which will have serious implications for keeping your animals healthy and fertile (see Chapter Eleven). The importance of soil testing cannot be emphasised enough. Talk to your neighbours and ask for their advice. Some fertiliser companies offer a very cost-effective testing service in the hope that you will eventually purchase their products, but you should not be under any obligation, take advantage of their generous terms.

Vets and Friends

If this is your first livestock venture, it will be important to find a good local vet, who will treat you with as much respect as one of his or her larger customers. You need a health plan for your animals, and routine worming and vaccination may be necessary depending on the circumstances. All of these can be discussed with your vet. You may well also need help with calving and other routine tasks, at least until you have built up your experience. Neighbours may be happy to assist and will welcome your help when they need a hand too.

Handling

There are times when you will need to handle the cattle in a controlled way such as for the bi-annual herd test for TB and Brucellosis (see page 95) or for other routine tasks. The ideal option for this is a purpose-built cattle crush, although these are expensive. You could make your own however using some 8 ft or 10 ft gates which can be acquired quite cheaply, if you can find some second-hand at a farm sale. Less expensive non-galvanised gates would also be acceptable.

Study

It is a good idea to read widely on the subject of cattle management and health. When looking at cattle health books however bear in mind that it is unlikely that you will encounter the diseases and disorders covered if you look after your cattle well and follow advice. A useful Further Reading section appears at the end of this book.

Red Tape

You will need to keep an up-to-date herd record and a Medicine and Movement book (this is a legal requirement, your local Animal Health Department and DEFRA will advise on how these should be kept and they will be inspected from time to time). This information will also be useful to you for management purposes. You should also be aware of the strict tagging identification regulations and welfare rules for cattle (details from DEFRA and the British Cattle Movement Service (BCMS)). Tagging is important for traceability, as highlighted by the UK's recent Foot and Mouth disease outbreak. By identifying each individual animal, which you need to do for pedigree work as a matter of course, you are also providing an audit trail from field to plate for the consumer. Tagging also provides the ministry with the means of tracking and isolating contagious disease outbreaks.

Subsidies

In the UK, at the time of writing, you may apply for subsidy quota and you can also lease extra quota and there is a headage payment for suckler cows and beef and bull premiums – contact DEFRA for the necessary information, however there is talk of such support being removed so don't base your enterprise on subsidies. Look upon it as a bonus while you can get it. While it is likely that agricultural policy will be reformed at European level, it may be that traditional breeds will be favoured for future subsidies.

Cattle Passports

In the UK all animals must have a cattle passport (administrated by BCMS). This document must always accompany the animal when it is moved. You should take care not to lose or damage it, as they are expensive to replace. For your own benefit keep a separate file on each animal for your own management use.

Insurance

If you already keep stock you probably have some insurance arrangements, if so, you should cover your cattle for straying and perhaps death by lightning if they are reasonably valuable. 'All Risks' cover is expensive and it may be difficult to justify the cost. If you need to start a policy you *must* have Public Liability insurance in event that your stock damages vehicles, other property or, worse still, people. If you are likely to have visitors it may also be worth checking that the insurance policy covers animal diseases which can be caught by humans (called collectively zoonoses) such as E. Coli, Salmonella etc.. Fortunately tailor-made insurance policies for smaller enterprises are available, some brokers advertise in specialist magazines.

Record Keeping

There is a certain amount of record keeping and paperwork involved with keeping cattle but do not let this put you off. If you are going to keep pedigree animals most of the information will be useful to you anyway as you will want properly identified individuals and a record of their breeding history etc..

Buying animals

Non–pedigree and unregistered stock

It will undoubtedly be cheaper, in the short term, to acquire crossbred or unregistered pure-bred animals if your only interest is simply meat and/or milk. These animals can be purchased from a livestock market but there are risks involved in this.

If you purchase from a market you may be buying an animal which has something wrong with it and you may not have any idea of its potential for your requirements, even if you only want a young animal you simply wish to rear for the freezer.

You may import disease and by the time you discover this you will not have any guarantee to fall back on (you usually have to notify the auctioneers within 24 hours of purchase if you have a complaint).

You will have no true measure of its temperament.

The best option when buying non-pedigree or unregistered purebred stock is to buy from a neighbour, who can tell you something about how the animal was bred, what you can hope to achieve with it and by what means. You have also minimised the possibility of importing disease as you have dealt farm to farm. Hopefully your neighbour will also be prepared to give you advice until you have found your feet.

Pedigree Stock

Pedigree stock has several advantages over other cattle.

- If the animals are of high enough quality, the breed name will provide you with a unique selling point.

- The females may have a higher value as breeding animals than as meat.

- Occasionally bulls can be sold for breeding, but do not make the common mistake of thinking that every bull calf can go this way, demand is low and quality should be exceptional, regard such sales as a bonus with most breeds.

Traditional Hereford family group showing typical cattle behaviour!

- Many small-scale breeders were insulated from the effect on meat prices of the BSE crisis in the UK because their animals could still find a market for breeding.

- Pedigree animals have been selected carefully over many generations so the breeder you acquire your foundation animals from should be able to give you detailed advice on how to rear and finish them in the most efficient way, and are usually more than willing to do so.

- Pedigree stock generally costs more than unregistered animals but they do come with a warranty as defined by the individual breed society, ask the breeder for details. The only other additional cost to choosing pedigree animals is the annual membership of the breed society and registration fees, which vary between breeds. In this context a lot will depend on the size of your enterprise, but any extra costs over non-pedigree or unregistered purebred stock should be covered by the premium prices you can aim to achieve for your end product (See Chapter Thirteen on marketing).

What animals should you buy to begin with?

The most cost-effective ways to start are as follows.

- If you simply want to produce some beef for yourself, then buy a couple of weaned steers, bull calves which have been castrated, as this will save you the trouble and expense of calling out the vet.

- If you are going to breed you could buy weaned heifer calves, but they will not be ready to put in-calf until they are 18 – 24 months old.

- The next least costly option are 'bulling heifers'. These are ready to breed from at something like 18 – 24 months old, but you will then have to go to the trouble of getting them in calf.

- In-calf heifers are the next option but you will need to check that they are in-calf by getting your vet to carry out a pregnancy diagnosis (PD) by rectal palpation.

- The best option, if you can afford it, is to buy older in-calf cows with calves at foot. The cows will be 'old hands' and you can learn a lot by watching them and by seeing how their existing calves develop.

Ask about an animal's breeding history before buying to ensure she has a good record. Cows can continue to breed up to around 18 or 20 years old but the majority, for various reasons, do not. A friend's 20-year-old Hereford produced 18 calves (see Figure below) however this is unusual. Generally any calves produced after 12 years old should be regarded as a bonus. You may find that individual animals, particularly of the traditional breeds, will greatly exceed your expectations.

Longevity: David Powell's 20-yr-old dehorned Hereford cow with her eighteenth calf.

Chapter Five
Management and Facilities

Tall hedges and trees keep the animals' backs drier in the rain, and provide shade from the sun.

Management

Cattle generally require less in the way of management than other farm livestock. For most of the year they are content in the field, and therefore the demands on both the animal and the stockperson are minimal. Cattle are best kept at grass for as much of the year as possible, but do not assume that any old patch of ground will do, to keep them happy and healthy they prefer some shade when it is hot, and some form of shelter particularly in prolonged wet periods or as protection against lightning.

Shelter

It is a good idea to let hedges and trees thicken and overhang fencing around your field, this benefits both wild species and livestock. The cattle can defend themselves better against flies, since flies cannot congregate in the space between the animals and the branches and the animal scratching against the branches also helps keep them off. Tall hedges and trees keep the animals' backs drier in the rain, and provide shade from the sun. They also enjoy browsing the leaves and no doubt acquire some useful minerals by this process. In the winter these hedges act as a windbreak. You will need to carry out some trimming from time to time, particularly if a hedge is thinning at the bottom creating a potential escape route.

Through judicious use of gates held together with rope it is possible to make most farm buildings accessible to your stock during periods of bad weather. However it is still a good idea to provide a basic field shelter or two, and this can be done without breaking the bank. The size will of course depend on the numbers of animals you have. You can use many kinds of cast-off materials including doors, pallets, sections of pre-cast concrete cladding, dismantled packing case materials etc.. A shelter needs a back, a roof and two sides, it should be built on a well drained area and you may need to put down some stone, hard-core or other material at the front to avoid a mudbath in the worst weather. It is also possible to acquire batches of bagged cement which is close to or just past its sell by date if you want to do a 'proper job' of the base and frontage. Take account of the prevailing winds when deciding which way the open front should face. Once the shelter is built you should make sure the floor area has some straw or other litter to provide a comfortable lying area.

Prefabricated field shelters are available, they are not cheap but the suppliers will erect them for you. It can be a good idea to obtain brochures and study the designs before embarking on a home-made version.

Housing

It is better for your cattle to stay out for as much of the year as possible (and it also means less work for you) but at some point during the winter you will have to bring them in to save too much damage to your pasture, as well as for their own welfare since being up to their bellies in mud is not good for them or their calves. Cattle being managed in a ranching system will of course not be offered these benefits – here the 'survival of the fittest' will apply.

The cattle will not require too much input from you to be comfortable and content while housed, but it is quite hard work to maintain the standards required for them to be kept clean and healthy if being housed for any period of time. The plus side is it will help to keep you fit and motivated through a long grey winter!

It was traditional in some parts of the UK to confine a cow by a neck chain or halter to an individual byre or standing. This makes the management easier for the stockman and saves space, but it prevents the animal from having its normal freedom of movement and all that this implies. In some locations this system has been used for centuries and is still employed through necessity in particularly inhospitable environments, however, and with the best welfare standards for cattle in mind there are a number of other options to consider.

If you are planning to build your own housing or have it built professionally, you should research the buildings available commercially to get some idea of current designs. Books with designs can also be ordered from the library. If on the other hand, you already own some farm buildings, they can be adapted for your new purpose cheaply and easily.

Housing should consist of a roofed area to keep off the rain, and walls from blocks or bricks to about five feet tall for shelter from the wind. The higher part of the wall can be made of less robust materials, spaced boards (Yorkshire boards) are commonly used as they aid ventilation but still keep out the worst of the weather. It is also a good idea if you have the space to have an open yard area for the animals to exercise in, the floor of which should be solid and well draining.

The height of the water trough needs to take account of the fact that the floor level will rise as you add fresh straw bedding over time. The front of the housing area must be kept clear for access for feeding, adding bedding, mucking out and the movement of an animal that may need to be isolated because of illness, injury or calving. Space requirements will vary considerably depending on breed and whether they are horned or polled and if horned, how wide the horns are. You should allow as much as seven feet of trough space each for the widest horned cattle such as Highland or White Park, but this would be excessive for those with shorter horns.

In-wintered cattle need:

• as big an area as possible, with plenty of space to lie down;

• as much fresh air as possible, but with protection from strong draughts;

• as much light as possible;

- regular clean bedding for comfort and to help them keep themselves clean;
- a constant supply of fresh clean water – regularly check the trough for fouling;
- sufficient clean trough space for any hard feed you may give.

Cows can become bored, so ensure they can see the daily goings on on the holding or farm from their accommodation to give them some interest.

You will need to allow plenty of room for access to ring feeders for silage or feed straw to avoid jostling and so minimising potential accidents.

Ensure that you have the appropriate number of feeders and sufficient trough space for the number of cattle you have, otherwise injuries are likely.

Feeding barrier suitable for horned cattle.

If you have horned cattle, ensure you use only properly adapted equipment to avoid injury (feeders, barriers etc.).

Isolation Facility

You may have to separate an animal from the main group because of illness or calving, this can be a warm dry loosebox, a stable is often used for this purpose. DEFRA will expect you to have such a facility if you intend to keep cattle.

It can also be useful to have an extra space in the event that you have one or even a few animals that you would like to treat slightly differently to the rest. You may wish to give them more or less feed than the other animals, they may for example have small calves at foot who would be better off being away from the main group of animals. Having an isolation facility can give you extra control and will make a big difference to your results.

For specific information on the housing of calves see Chapter Ten, page 82.

Bull Housing

If you keep bulls year round, housing them can be one of your more expensive undertakings. Some breeders keep bulls outside throughout the year on a 'ranch' system, but it is far simpler, if you can, to get a bull to run with the herd as and when you need one and preferably when the herd is in the field. There may however be times when you want to get cows in-calf during the winter months. If the area you keep your main group of cows is large enough, the bull can stay

Feeding barrier suitable for polled or dehorned cattle. Note movable troughs.

with the cows in the normal way, but you will have to remove any young heifers or bulls of six months old and over to prevent problems. You should also be aware of the potential for accidents when he 'performs'.

Dealing with Bulls

Good housing is particularly important in the case of bulls, as a well-reared bull is less likely to be aggressive than one which has been confined to a dark pen, has been neglected and has suffered sensory deprivation. There are exceptions to this of course and there is no room for complacency whatsoever. The safety issues of bull rearing should not be exaggerated (it has been an extremely enjoyable and satisfying part of the job for me), however you must not allow a momentary lapse of attention ruin an important part of developing your herd by natural service. At worst the results of too much trust or poor welfare can mean serious injury or even death to you or others. If you follow some basic rules this does not need to happen.

- Never turn your back on your bull.
- Always carry a stick, not to beat him with, but as an arm extension if you are having trouble with him and you need time to make an exit.
- Remember that his pen is his territory, and he may feel like reminding you of that when you least expect it.
- A bull needs to be able to see what is going on.
- He must not be able to escape, injuring himself or other in the process.

It is worth remembering that the animal you trust the most is potentially the most dangerous animal on your farm, and remember this can apply equally to cows!

Traditional Hereford bull. Photo courtesy of Rare Breeds Survival Trust

The Bull Pen

There will be times when it is preferable for the bull to be isolated and the following specifications would provide good accommodation for him as well as a proper level of safety for you.

The ideal bull pen consists of a roofed bedding area filled in on three sides, attached to a generous open exercise area. Construction is usually of brick, if using cement blocks care must be taken to ensure that the construction is of sufficient strength. The walls of the exercise area must be high enough to contain the bull, but low enough for him to be able to have a view of the daily activities on the farm. To prevent him escaping over the wall, inset two parallel steel rails on any vulnerable side(s).

A very useful addition is a mechanical 'yoke' set into the wall, similar to that on a cattle crush, with a feed trough on the outside of the pen. He will soon become used to being yoked for feeding, and it means you can enter the pen safely to do routine tasks, including some jobs on the bull himself.

Another excellent option is to build a metal escape door into one of the walls which you can leave open while you are in the pen. The door will be too small for him, but you can quickly escape through it in an emergency.

These two items will probably need to be purchased from a specialist company or contact your local blacksmith.

On a personal note, the bull pen we have is less than perfect. It consists of a roofed area, with a spacious yard and two gated access points. The current occupant has the worst temperament of any male we have reared so far. He gets his food before I enter and I am particularly wary of him whilst working in there, at all times I am aware of my best escape route and I have a stick to hand. I never give him the chance to take any liberties, and you will probably not be surprised to know that he is to be consigned to the freezer. Bulls from traditional beef breeds are usually docile if well managed and reared, whereas dairy breed bulls have a bad reputation and should not be taken on by the inexperienced. This is a generalisation, and as I have said before, individuals vary. Animals of unsound temperament should not be bred from, however good they look in the flesh, or on paper, it is far better to have a nice quiet herd that is a joy to work with.

Moving Cattle

Calves

When a calf is small, it may be picked up with one arm under its neck, taking care not to constrict the windpipe, placing your other arm over the animal's back with your hand under the flank. However it will not be long before they become too heavy for this, and they can kick and struggle in the process of carrying them. If you have the time, halter training is an option but it is unlikely to be practical to train a whole batch of calves. Once they are used to being fed by you, with a little help from someone behind to urge them on gently, it is just as likely you can get them to follow a bucket of feed. Some people even place a finger in a calf's mouth so that it thinks it is feeding time.

Prepare the route in advance by making raceways where possible with gates or sheep hurdles, making sure they are stable and cannot fall over. If you are patient the whole exercise will be painless for man and beast. Always make sure any gates to the road, fields etc. are closed in advance of any stock movements, so that any escapees cannot go far and you can avoid unnecessary stress to yourself and danger to the animals and road users.

Halter Training

Halter training takes time and training the animals to walk on a halter takes even longer. If you are planning to show, you will already be prepared to commit the time required, if not, it is still worth taking the time to acquaint your permanent stock with being haltered for management purposes, particularly if you do not have a crush for routine tasks like worming, herd testing and so on. Ideally you should invest in a crush as even the least co-operative animal can be manoeuvred into one. It is possible to keep cattle without a crush but in this case the halter will be essential at times.

- You can make halters yourself or they are available in different sizes from your agricultural merchant, who will also advise on the correct sized halter. I would recommend cotton to start with as they are less likely to cause rope burn to you and are kinder on the struggling beast. They may be easier to untie/release than polypropylene. If you prefer you may use polypropylene later once you and the beast are more experienced.

- There are many theories on how to achieve successful halter training. I have found the following method successful, and the least stressful for the animal.

- The younger the animal is when you start, the easier it will be.

- Get used to how the halter works, it is simple to use but it can be easy to get the loose rein on the wrong side if you are inexperienced.

- Never wrap the lead rope around your hand, always hold it in your palm, and use both hands to hold it if necessary for greater control.

1. Start with one animal in a confined space, preferably with a helper. Keep to the side of the animal for safety. Place a bucket of feed close to where you intend to tie the animal up. Place the large loop over the horns and ears, then the smaller loop over the snout, ensuring that the slip rope is under the jaw.

2. Tie a knot under the jaw so that the sliprope under the jaw cannot go too tight, this is particularly important with youngstock. It may take several attempts to get this right, since if you tie the knot in advance the fit may well be too tight or too loose, so it may take several attempts to get this right.

3. Get your helper to lift the feed bucket up so that the animal's head is off the ground and fasten the loose end of the halter to a strong point by a slipknot (see Figure on right). As soon as the bucket is removed the animal is likely to lean back with its full weight to test your knot tying ability!

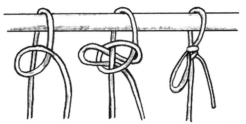

How to tie a slipknot.

4. Always use a slipknot or quick release knot to fasten the animal so that they can be released quickly and easily in an emergency.

Congratulations, you now have a reluctant but restrained animal on a piece of rope!

5. The next stage can be quite alarming but fear not, the animal may roll on its side or onto its back, its eyes may roll, or most alarming it will pretend to be dead or it may try anything it can to escape. Remember, the animal has never experienced anything like this before. All the way through this process you should be reassuring the animal and re-presenting the food.

Eventually most animals can be stood up again, even if somewhat reluctantly. You should stay with the animal and encourage it to feed again, most will.

6   Lesson one is nearly over. Once the animal is feeding again, retire to a point where the animal is alone but you can still observe in case of mishaps. The whole operation should hopefully take around a half to three-quarters of an hour.

7   Return to the animal which by now should have calmed down and carefully release your slipknot, while still keeping some food before it. If possible tie the lead rope around the halter so the animal can be released with the halter still in place, the halter can then be left on the animal to become used to. Make sure there are no hazards which the animal could get caught up on.

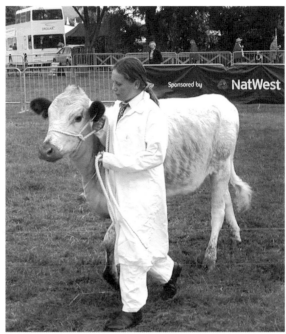

Well turned out participant in young handlers competition, with Irish Moiled heifer.

8   Repeat the tying up with feed exercise daily for a few days, for half an hour each time, ensuring that the animal relates the experience to having a nice feed rather than having needles plunged into them, hoof trimming etc., which is of course what you will be doing in future! Check the animal regularly.

As mentioned previously there are harsher ways to halter break an animal, but I believe the above method is kinder and well within the scope of the small herd keeper.

Training to walk with a halter is covered in Chapter Twelve.

Separation

At times animals need to be separated, such as calves at weaning, injured animals, or those you wish to put on a different feed regime to the others. Do not assume however that if you move a cow and her calf to an adjacent field that she will happily stay there. She is likely to make every attempt to rejoin the herd, even if she can rub noses with her friend over the fence or gate. I have seen mature cows attempt to leap gates and fences, causing significant damage to the fittings and worse still to themselves. If you have to separate stock from the herd, go to extreme measures to ensure that however hard they try, there is no way back to the herd and that they cannot damage themselves, even if this means housing them securely for a period. If you do house them, do not assume that your normal set up will contain them. We have resorted to tying sheep hurdles to the tops of normal gates and raising the barrier level all around a pen to keep an animal safely in her pen. If the animal can be kept out of sight and earshot of the herd even better, but this is rarely possible on a small place. This may seem extreme, but in my experience it is the only way. An empty bull pen or isolation box can be really useful at these times.

Basic Equipment List for the management and movement of stock

Small trailer for calves (and sheep). Note that moving stock in the back of a 4x4 or van is not allowed unless the vehicle has been specially adapted with a ramp. A small trailer is also useful for transporting feed and small quantities of hay as well as for feed collection.

Large cow trailer or **horse box** depending on the size of your enterprise. Alternatively arrange to borrow or hire from a neighbour, or use a professional haulier for occasional movements on and off farm.

Ring feeders can save a lot of labour particularly if you are using round bales of silage or feed straw. There are numerous suppliers of specialist equipment, who can provide you with their catalogues. You will need a special or adapted type for horned cattle.

Feed barriers are a 'stockman friendly' way of providing hard feed.

Strong gates for constructing instant penning, raceways, and basic handling facilities.

Cattle crush

Cattle crush (see Figure on right) this does not need to be expensive or 'state-of-the-art'. If you cannot find one straight away, keep looking and use gates to construct a handling pen when you need one.

Tractor, again this does not need to be expensive but consider safety aspects carefully.

Chain harrow. – this is essential for good pasture management.

Selection of items such as muck forks, buckets, sledge hammer, post rammer

Halters

Safety first: A Cautionary Tale

A breeder of Beef Shorthorns, in his seventies, was retiring. He had grown up with cattle, had spent his whole life with them, and had made his living from them. He was looking forward to a long and hard earned retirement in Canada. His pedigree herd was put up for sale but he decided take his favourite bull with him. Once settled in Canada, he was talking to his bull in the pen, as he had always done, when the bull attacked and killed him without warning. This type of unpredictable behaviour is also possible with a cow, particularly if she has a calf at foot. You should always be on your guard however well you know your animals.

Health and Safety – Tractors

Big bales are becoming increasingly common and many people do not have the correct handling equipment to move them. If you are using a classic tractor to move big bales, will you be safe if a bale rolls down the loading forks on to the tractor? If it does not have a cab, I doubt it. Bear in mind that you may not need to buy a more suitable tractor as it may be possible to modify it quite cheaply with professional help. Or you may be able to borrow a safer tractor when you need to move big bales.

Chapter Six
Feeding

The quality of any home-produced forage rations will depend on the weather during the growing season and the time of year the forage was cut and this will affect protein levels and dry matter content. For example, if the weather was very wet in the lead up to cutting, quality will be affected. The same is true of other feed crops including grain. It is possible to have your crop analysed to find out precise protein and dry matter levels.

Initially you may wish to take a scientific approach to feeding by referring to some of the titles in the Further Reading section, however as you gain experience you will be able rely on your own judgement.

Feed requirements will also vary with the weather, type of breed, whether or not the animals are housed, as well as what you are expecting from them, i.e. whether you are simply maintaining them in readiness for breeding or whether you are aiming for meat/dairy production. In addition the availability and quality of fodder and straw will vary from year to year.

Feeding regime (December–April)

Let us look at managing a small suckler herd in the winter. We should start with the obvious fact that cows eat grass, therefore conserved grass is traditionally their mainstay for the winter. However, as the quality of conserved grass can vary tremendously your skill and judgement will be required to balance out your herd's needs.

Condition

When considering feed requirements, you must remember that your herd consists of individuals and they will all need to be assessed as such. For example, a young cow with her first calf will struggle to 'do' her offspring well if on survival rations alone, even if she 'milks off her back'. This loss of condition tends to be less of a problem with mature cows and it may not be of any great concern if the calf is not of any great value to you, but if you plan to include it as a future breeding animal, or wish to 'finish' it in good time, it may be worth separating the calf and its young mother from the rest of the herd so that you can manage them differently. In another example, the boss cow may be bordering on fat after a good summer, and as a mature animal may have plenty of milk for her calf, so you can afford to be a little harder on her.

You may simply keep your animals on a 'ranching' system, where they are over-wintered outside with minimal food supplementation. Remember that if they are kept too hungry they will do all they can to leave your place in search of food. On the whole, if you keep the animals from being hungry, even with straw or poor quality hay, they will survive quite happily. You may wish to treat them a little better towards the end of winter if your budget allows. It is worth bearing in mind that cattle will not readily go back to poor fodder if you suddenly give them sweet silage or high quality hay. Therefore save the best till last and be sure you can sustain the better feed until the grass grows. Please note: it is very important that changes to the feeding regime should be introduced gradually to avoid causing health problems in your cattle (see Chapter Eleven on health).

The main thing is to keep your animal's stomachs reasonably full in order to keep their gut working properly. You will probably need to provide some form of supplementary feed whichever feed regime you adopt – the simplest form of supplementation being feed blocks which contain vitamins as well. (See following for detailed information on types of supplementary feed.)

Condition scoring

There is a recognised, 'condition scoring' system mostly used by commercial producers which may be of help, but most livestock keepers use their eyes. A basic way of assessing the condition of your cows is to look at them from behind. If they are 'broad in the beam' and you cannot see their hip (pin) bones the likelihood is that they are too fat. If on the other hand the bones protrude the animal would be classed as being in poor condition – it is best to aim for a happy medium. You should note that overweight animals are less inclined to breed well, so any overweight cows you may have should be placed on a harsher regime, this applies both to mature cows and young heifers.

Types of supplementary feed

Hay

The degree of your self-sufficiency in fodder will depend on your herd size compared to your available land. You may decide not to make your own hay in favour of being able to keep a greater number of animals particularly if pedigree breeding stock is your main enterprise. There are pros and cons to this approach.

Pros

- You will need to keep only a small amount of machinery: a small tractor, a chain harrow, a buck rake and a tipping flatbed trailer, which will save you space and money.
- You will not have to plan your life around haymaking weather.
- You can choose where to buy fodder according to price and quality.
- You can buy hay 'from the field' as it is made, and get it at the best possible price, by loading and transporting it yourself.

Winter feeding.

Cons

- If it is a poor season, you pay a higher price for hay which may be of poor quality.
- As small bales are becoming more difficult to obtain you may need to be equipped to be able to handle large round bales or the even heavier silage bales.

With the more traditional breeds, they could, on a simple maintenance-only regime, be expected to survive the winter on good to medium quality hay alone (save for a molassed mineral lick for energy and vitamins, see page 54). However, you may find that overall hay quality is poor and that they need feed supplementation, either concentrate (hard feed)or feed blocks (see below).

Warning

Although animals can be fed poor quality hay, never feed mouldy hay. The mould spores from this can be harmful and can cause respiratory diseases if inhaled. If you open a bale that contains mouldy sections then commit those parts to the muckheap.

Straw

Straw can be the main bulk ration throughout the winter, particularly for traditional breeds. Barley straw is the preferred option, although it should be noted that its nutritional value has decreased over recent decades. Some cattle may at times show a preference for wheat straw but nutritionally it may not be as good as barley straw. Use feed blocks or sugar beet (pellets or pulp) as a supplement.

Warning

As with hay, mouldy straw should be avoided as the mould can cause respiratory problems if inhaled. Although the animals will tend to avoid the mouldy parts, unless they are really hungry, because it tastes sour, it is better to exclude the worst .

Silage

Silage can be described simply as pickled grass and it is becoming the most popular form of forage conservation. It is less dependent on the weather for its production. Some larger enterprises make huge silage 'clamps' where the cut grass is layered and then covered in a purpose-made area, but the wrapped round bale is the mainstay of most operations. If you do grow your

Big bale silage.

own grass, then the most obvious option at harvest is to employ a contractor with the specialised equipment needed to make wrapped bales. It can be difficult to get a contractor to agree to cut your small acreage precisely when you want them due to demands from their larger customers.

Wrapped round bales weigh around half a tonne, so you need equipment which can handle them – if you lose control of a bale on your front loader it could easily cause serious injury at the very least. Make sure your machinery is safely up to the job.

Note that once stacked these round bales should only be moved when needed – if you puncture the plastic wrap the silage will spoil. Once opened the silage needs to be used within a day or two.

Cattle love sweet succulent silage and do very well on it. Silage does have a laxative effect, adding a small amount of straw to their feed will help lessen this.

Warning

Again, you will need to remove any mouldy sections from your silage for the same reasons as with hay. It is also worth bearing in mind that the liquid 'run off' or effluent from stored silage is a pollutant, arrange your stack away from water courses, wells, streams etc.. (Legislation concerning silage effluent is becoming increasingly strict.)

Haylage

As the name suggests, haylage is halfway between hay and silage. For haylage the grass is left to dry a little before baling, unlike silage where the grass is baled very soon after it is cut. Haylage is therefore drier than silage and usually has a higher protein content. This is an excellent feed if you can produce it, however it is a popular feed for horses, and is therefore usually more expensive to buy. Please note that as with hay and silage any mouldy haylage should not be given to your animals to avoid potential health problems.

Feed blocks

For the upland herd or those run outside, feed blocks are an excellent option. They usually weigh 25 kg and in the UK range in price from £6.50 to £8.80. Savings can be made by buying in bulk but how cost effective this is will depend on the scale of your enterprise. It is also possible to buy larger 'feed buckets' but these usually are delivered in minimum quantities.

Different formulations are available, such as feed block with a high magnesium content. This type of feed block is vital when the grass starts to disappear and your cows may be susceptible to deficiency problems (see page 90). With the help of blocks you can outwinter your cattle, on poor quality fodder. They can help themselves to the block as and when they need it.

Depending on conditions, a block can last five cows, a mature bull and following calves up to six days, making it a cost effective option.

'Commercial' feed blocks are not allowed for organic production, but I understand that blocks which have been approved for organic production are now available. Ask your organic certification body for details of suppliers.

Concentrates (hard feed)

If you are fattening cattle for meat you will need to use some hard feed in the winter. You may also prefer to use hard feed for the whole herd if the bulk fodder is poor or you are feeding straw. The simplest way is to purchase proprietary feeds is from your local mill or supplier. I was taught to be cautious about this approach by a great stockbreeder, indeed, he used to say to his customers, 'On no account should stock from this farm be fed on pelleted or powdered mysteries in bags'.

There is still no law requiring a feed compounder to state exactly what their feeds are made from, they regard it as a trade secret. In the UK you can at least be sure that animal protein is now excluded from these 'commercial' rations following the outbreak of BSE during the 1990s. There is also an increasing number of specialist feed companies which produce rations made from GM-free ingredients or rations are described as 'additive free'; prepared organic feeds are also available. Of course all specialised feeds cost more.

The advantage of buying any proprietary feed is that it is 'balanced'. That is to say it has specific amounts of the right nutrients, vitamins and minerals to ensure that you are supplying all your animals needs. A proprietary feed with a protein content of between 14 and 16 per cent is ideal. You can specify protein content when you buy ready-made feeds.

For suckler herds, 'Beef Nuts' (commercially produced feed pellets) are the most common type of proprietary feed used. Feed merchants also produce what is known as a 'Coarse Ration', this is the cattle equivalent of muesli, as opposed to the compounded 'nuts', and the cattle much prefer it. It is often used by pedigree breeders for their most important animals. You are likely to pay an extra £2 or more per bag for this feed.

Home-mixed ration.

Own mix rations

Another option is to mix your own rations from 'straights' – individual ingredients such as cereals, soya or peas and beans for protein, distiller's grains, molasses, dried sugar beet etc.. You will need a mill for the corn and mixing your own rations is time consuming, but the advantage is that you know exactly what your animals have had, it will always have been freshly prepared and they will do well on it. Note that you will need to include a mineral and vitamin supplement. Please note that as a result of BSE you need to register as a home feed mixer, with your local DEFRA, no formal licence required but you MUST be registered.

One of the disadvantages of own-mix rations is that you are unlikely to be able to grow enough corn or other fodder crops to be self sufficient, and so you will have to take into account any transport costs when buying in feed and have a suitable storage and mixing space. It will not necessarily be the cheapest ration but it can be the best.

It may be that your neighbours will be prepared to sell you some of their own mix. They may have the economies of scale to be producing a ration for themselves and have the necessary permission from DEFRA to supply you. An increasing number of farmers are mixing for themselves – it is a good way for them to 'add value' to their grain, especially during years where grain prices are low.

There are many different 'recipes' for own-mix rations and there are lots variables to be taken into account, including the exact nutritional value and quality of the ingredients available and so on. You will find various mixes in the more technical books listed for further reading

A simple own-mix ration, is to use rolled barley as a first choice, or a mixture of barley and another corn variety, to which you could add some commercial 'nuts'. By adding the nuts you will be ensuring that vitamins and minerals over and above your simple grain base will be included in the cheapest, freshest and most convenient way. It may be possible to acquire some corn locally at a good price, but you will need to roll or grind it as the whole grain cannot be readily digested, a lot of it will pass straight through the animal. Please note that barley should be introduced very gradually as it can kill if given in too large quantities too quickly. This is known as Barley Poisoning.

Cheap concentrates

It is possible to buy bulk loads of low cost concentrate ingredients and mixes. In my area there is a ration available that is made from crisp (potato chip) waste, distillers grains, pot ale syrup and minerals. It smells delicious, the animals love it, but it goes 'off' fairly quickly so you have to use it up efficiently. It is at least £1 cheaper per 25 kg than other rations.

In some parts of the UK you can get what is called 'sweet mix' made up from a variety of sweet and biscuit wastes. It is a balanced ration and the animals will do well on it, but its main disadvantage is that vermin love it too. Another point to watch for when buying in any kind of commercial ingredients is how fresh they are, do not think that commodities past their sell by date are a bargain, they could be sour and potentially harmful.

Roots and cut-and-carry fodder

This is the most traditional way of getting cattle through the winter.

Fodder beet can be given whole, it is an excellent feed, but it is possible for cattle to choke on root crops so it may be best to chop it into chunks. You can buy stock-feed carrots, mangolds, potatoes and swedes, but you will have to order a minimum load, and ensure that they are all used up before they spoil. Beet tops, kale and cabbages can also be grown as a winter feed. Note that beet tops must be wilted before being offered, and too much kale and cabbage can have an adverse effect on fertility. By all means offer any surplus from the vegetable plot, but they may turn their noses up at things they are not used to. Stored apples which are surplus to your requirements or are not keeping too well (not cookers) are a welcome treat, but again do not overdo it.

From experience, I would say that preparing fodder is hard work. Chopping frosty fodder beet on a cold winter's morning is not my idea of fun and it is very labour intensive.

Vitamins and minerals

If your holding has any vitamin or mineral deficiencies, then these will be carried through into any home-produced fodder. Although not allowed under organic rules, where possible, animals should have access to mineral licks throughout the year, molassed in the winter for extra energy, or the salt type in the summer. Another course of action as a 'cover all' is to administer a multivitamin and trace-element bolus (a bolus is a large, slow release 'pill' which is swallowed by the animal). Discuss this option with your vet. Ensuring that you have done all you can to prevent any shortages of essential vitamins and minerals will mean that your cattle will be healthier and better equipped to cope with an arduous winter.

Feeding young stock

It may be that you want to help some of your younger animals along without feeding their mothers. In such cases you can use a 'creep feeder'. These are available commercially, but can be expensive for the smaller enterprise. It is possible to construct a creep area for very little outlay.

Feeding training youngstock go well together

If housed, section off an adequate space, with a narrow access point through which only the calves can enter, allow plenty of room for them. Some people make creep feed available on an ad-lib basis, but this can attract vermin, and can be expensive. The calves will soon get used to a routine of being fed in there, for example, twice a day. The creep needs to be well constructed and strong enough to prevent the envious cows from trying to barge their way in. You can also construct such areas in the field using posts, netting and barbed wire. Your main difficulty will be refilling the troughs without attracting the attention of the adults, so this is best carried out after the adults have started feeding. Another option is to site the creep area near the yard so that it can be topped up from the outside. Giving your calves this special treatment also means that once spring arrives they will already have a positive view of humans, making halter training and other handling tasks easier.

Finishing stock

By finishing we mean bringing an animal to peak condition for killing. An animal said to have a good 'finish' is one that has put on flesh in all the right places and has just the right amount of fat under the skin to help to ensure that it will cook beautifully. If you are finishing the animal for yourself, you need not be quite so particular, but if it is destined for a butcher he will have precise standards for you to aspire to. You could do worse than ask the butcher to look over your beasts before you book them in for slaughter.

Finishing stock for meat production is a real art, but with experience it does become easier. Below are a few basic tips. For greater detail see Further Reading section.

It is better to fatten two animals together rather than one on its own. Two animals also travel to the abattoir less stressfully and the result is better meat.

If you intend to sell a finished male animal to a butcher it is best for it to have been castrated. Most butchers will not accept entire males as beef from an entire male is sometimes considered to be of poorer quality than from a steer, although this is still the subject of some debate. It is also worth considering that a steer is likely to be calmer than an entire male which will help with management.

Do not try to 'finish' the animals too quickly as this will affect the meat quality. It is possible for a Traditional Hereford to be ready for slaughter at 14 months, but the meat will be far better at 18 – 24 months.

In the UK at the time of writing, all meat animals have to be killed before they are 30 months old (following rules imposed after the outbreak of BSE) this means that slower growing or larger breeds may need to be 'pushed along' to be ready within the time limit.

Depending on the breed, most traditional types should be around 500 kg to be worth killing, however if you are not selling to a butcher, the overall 'finish' and weight may not be so critical. If the animal has been kept on a non-intensive regime, and it is for your own consumption, getting every last ounce of potential from the animal may not be so important, the meat is still likely to be high quality if killed and prepared properly.

Your meat animals will need bulk feed such as hay, straw or silage, in the same way as your other adult stock, but their speed of growth can be controlled by careful use of concentrates. Animals are best 'pushed along' from around twelve months old, if you start your finishing regime any sooner the animals are likely to become too fat.

Giving too much concentrate will result in diarrhoea-like symptoms in the animals, if this occurs, immediately reduce the amount you are giving.

Two or even three smaller feeds per day are better than one large feed.

Feeding equipment

Cautionary notes

If you use ring feeders, make sure you keep them topped up. If they run too low, it is possible for a hungry animal to go to any lengths to reach the last little bit of feed, in doing so it may become trapped and injure itself.

Hay racks are the best way to ensure that expensive fodder does not get trampled into the ground, however, and this is particularly the case with dusty and 'stalky' fodder, racks can increase the possibility of eye injuries/infections. If using racks you will need to check your animals eyes carefully and regularly to pick up problems early on. If feeding on the ground, you can minimise wastage by only giving them enough to clear up quickly, you will soon be able to judge how much they need in this way. It is surprising how little is wasted by this method.

Keep any troughs for hard feed clean, and always check water troughs for fouling, even in the field, as troughs become fouled quite often. A clean water supply is essential at all times, particularly when feeding dry matter since with dry feed higher volumes of water are needed to aid digestion.

Commercial ring feeder adapted for use by horned cattle.

Conclusion

The winter months can be difficult for both you and your animals. However it will all be worth it! There is an intense feeling of relief and satisfaction when the grass starts to grow again, your animals are clean, their coats begin to shine once more, and they improve in condition. A suitable time to lean on the gate and admire the results of your hard work.

Feeding regime (May–December)

From May until around the end of December, depending on the weather, your cattle should be able to survive, breed and grow on grass if there are sufficient quantities of it. The nutritional value of the grass varies during the season, and much is dependent on the varieties of grass and other plants and herbs in your pasture. See Chapter Eleven, Health Care and Diseases for a discussion of the issues to be aware of during the grazing season.

Pasture

You should allow one acre per cow of reasonable quality pasture, slightly more if they have a calf, and this does not take account of any land you have set aside for hay or silage making. You will need to be more generous with poorer land and forage quantities will be greatly affected by the level at which you apply fertiliser or manure.

You should also pay attention to the varieties of grass which make up your pasture. Single-variety commercial pastures, or leys as they are known, are not the best option. These days most commercial pasture consists of hard-wearing rye grass, for highest efficiency of growth production and fodder conservation but on a smaller scale, you can take a wider view. Animals undoubtedly benefit from an increased variety of plants and this should be your long-term objective. Cows enjoy variety, and it is believed by many that if they have a choice of herbage they will graze selectively for their own good health. It is often argued that meat animals grazed on these mixed pastures produce better tasting meat.

In the UK there are a number of specialised seed merchants who offer a range of mixtures, containing a wide variety of plants and herbs as well as different grasses which will be suited to your land and benefit your stock. You can arrange for one of these seed merchants to come and assess the quality of your pasture (you are under no obligation to buy). The rep will identify the varieties you have and can provide expert advice on your particular circumstances.

It can be useful to include clovers in your pasture. Clover is known for fixing nitrogen in the soil and this can reduce the need for artificial fertilisers. Importantly, in certain forms, clover can improve fertility levels in your stock but this should be white clover with very small leaves, NOT the very leafy nitrogen-fixing red clovers used to increase pasture fertility (A word of warning – clovers can cause fatal health problems (see Frothy Bloat, page 92 for details).

This rich, varied grazing, developed over the longer term can be harder wearing than cropping mixtures which are designed for rapid growth and fibre content for conserved forage production. Some organic farms which have concentrated on developing varied pasture, look almost like salad bowls for cattle in the growing months, and the cattle appear to do very well on it.

Maintaining pasture

If you have a limited acreage the most useful item for pasture 'cleaning' and rejuvenation is a chain harrow. This can be used throughout the season as long as the ground is not too wet. It works by breaking up the hardened pats, scarifying and aerating the surface, this encourages strong re-growth allowing you to make the most of the pasture you do have.

Conservation grazing

In the UK, conservation grazing has gathered momentum over the last few years. The term is used when grazing animals, cattle, ponies, sheep or even pigs, are used to manage land in a scientifically responsible and productive way.

An early example of conservation grazing was the use of the small, primitive Soay Sheep breed by the English China Clay Company in Cornwall. The company produced huge spoil heaps as part of their operations and to landscape these potential eyesores the thin surface was sown with grass seed. Commercial sheep breeds would not have been able to graze the fragile sward without causing a great deal of damage and would have required a high level of management which would have made the project unsustainable. The small, nimble-footed Soay was perfect for the job as it is extremely self sufficient and hardy. A 'flying flock' was used with great success to carry out the landscaping work.

Similar pioneering work was carried out with another primitive sheep breed, the Hebridean, used to control the spread of a dominant coarse grass type called Molinea which was stifling rare and more delicate species on sand dunes in the north of England, which in turn supported wider species of insects and other wildlife.

Galloway cattle have always been known for their hardiness, and thanks to their thick hairy coats they have the ability to thrive rather than just survive in damp climates. Thus Galloways have been put to good use in places where other breeds would have struggled and larger heavier breeds would have destroyed wet pasture. Pigs have also been put to excellent use to reinvigorate overgrown poor quality woodland by clearing unwanted undergrowth.

In Germany, where vast swathes of land which would be difficult to manage sensitively and beneficially using modern 'commercial' breeds, both Galloways and English Longhorns have been doing a great job with minimal supervision. The by-products of these natural land management schemes are a supply of good meat, marketed as 'wild beef' and sold at a premium, as well as cost-effective breeding herds.

In the UK conservation organisations such as English Nature, who are responsible for many scientifically sensitive areas have started to use this approach. As part of the ongoing reform of the European Common Agricultural Policy (CAP) a Traditional Breeds Incentive scheme has been introduced for both lowland and upland areas, this will obviously encourage the use of traditional and native breeds. Some breeders who had been struggling to find more land have now had many more acres put at their disposal, under certain strict conditions, for at least part of the year and on preferential terms which has enabled them to improve their financial situation and expand their herds. In the UK , see Contacts section for organisations who can help. Outside the UK, contact your department of agriculture or non-government organisations specialising in land conservation. Conservation grazing initiatives have already had a wide-reaching effect and provide traditional breeds with their one of their brightest hopes for survival.

Conservation grazing in Germany using English Longhorns.

Chapter Seven
The house cow and dairying

Cheesemaking using Gloucester and Kerry milk at Wick Court Farm.

Once you have tried full cream milk, fresh from the cow, or home-made butter, you are spoiled for life – it tastes so very different from the store-bought product. Small-scale dairying, making your own cheese, butter, yoghurt, ice cream and so on, as well as home-produced milk, is about as satisfying as it can get for the smallholder whose main aim is self sufficiency.

Any surplus milk can be used to feed fattening pigs, or to rear bought-in calves for meat for yourself or to provide extra income, this can be given either from the bucket or by multi-suckling. Whey left over from cheese making can also go to your other stock, including chickens.

Legislation

If you intend to supply your produce to others you will need to check the legislation and regulations with your local authority, in the UK this will be Trading Standards and DEFRA. Dairy products are regarded as high-risk foods so your whole production process will be subject to scrutiny. However the importance of specialised local products is being recognised, and in the UK many local councils/DEFRA will be able to provide you with help and advice. It is also worth investigating whether any grants are available to help you establish your new business.

Dairying is a specialised subject and needs to be studied in detail. In the UK there are practical courses available, covering all aspects of dairying. In other countries you should seek out 'artisan' producers for advice if no courses are available. Below are some hints and tips to encourage you.

Choosing your dairy cow

Milking potential and temperament are inherited characteristics. If you are inexperienced, you will need a cow that will help rather than hinder, so enquire about the history of its relatives before you buy. Some breeders will be able to supply milk production records going back many generations. A placid and co-operative temperament is also important, at least until you are more sure of what you are doing.

Ideally, your first animal should be a cow which is used to being milked, but if you are starting with an in-calf heifer, the first step is to get her used to being on a halter and secondly, gradually to get her used to having her udder felt. Do this before she calves as part of your routine contact with her. The more direct contact you have with your animals, the easier all your tasks will be.

Hand milking

The cow's udder has four divided 'quarters' with a teat on each. If you watch a suckling calf you will see that it moves to different quarters whilst feeding, to achieve maximum output on a daily basis you need to 'strip out' all that you can, to ensure that the cow will produce a similar yield the next time. You also need to ensure that the udder is scrupulously clean, you can purchase 'udder wipes' specifically for this purpose.

Handmilking a Shetland cow on a smallholding in a built-up area in Lancashire. (Photo, Tony Bennett.)

If milking by hand rather than machine, do not use a stainless steel dairy bucket at first. If the cow is difficult and the bucket gets kicked over the noise it makes may upset the cow, getting you off to a bad start. Use a small plastic jug to begin with whilst you are perfecting your technique.

Milk the cow from the side according to whether you are right-or left-handed, in this way you can hang on to a teat and partly control any attempts by the cow to remove you with her back leg!

Do not become frustrated if at first the cow gives you little milk. The 'let down' reflex has a hormonal trigger. In order to trigger the 'let down' reflex the calf 'butts' the udder upwards as a demand sign, when hand milking, unless the cow relaxes and wishes to give milk, no amount of trying on your part will help. As a result she may not be able to give you much milk even if she wants to, she needs to get used to you. You can use the calf to help this take place until you have developed your skills by 'sharing' the cow with the calf, but do not use the milk for your own consumption if there has been a slobbering calf involved – in dairying hygiene is vitally important.

Talk to the cow while you are trying to milk her. Cows respond to kind words and tone of voice, if she decides to accept you as one of her 'calves' she will be delighted to produce more milk just for you.

Mechanised milking

The scale of your plans will determine whether to milk by hand or use a small machine. Specialised suppliers of dairy equipment to small producers are more common these days, and some can offer detailed advice on helping you get started.

Controlling milk output

It is important to note that the amount of milk produced can be controlled through your feeding regime. If you just want a little milk for the house, you won't need to offer much in the way of extra feed, but if it is quantity you want, and the cow has a calf to rear as well, you will need to step up her rations if she is to support your dairying requirements.

If you do not want to fully commit yourself to having a dairy enterprise, another way of controlling milk output is to use a cow's calf to moderate the amount of milk you take. Whilst the calf is suckling and the cow is eating, you can milk the cow from one of her free quarters, being careful to exclude any of the froth the calf is producing, and of course carefully clean the teat you are using with an udder wipe.

If you require larger amounts of milk, separate the calf from the cow for part of the day, having provided the calf with the appropriate hay, feed and water. Do not worry about any 'hooting and hollering', the cow and calf will become used to the routine relatively quickly.

If you only want a small amount of milk, you will not need a specialised dairy or dual-purpose breed. Your quietest beef cow won't mind you taking some milk if you get her used to the idea gently.

The health of the dairy cow

Dairy animals must be protected from specific health threats such as calcium deficiency (some breeds and individuals can be more susceptible than others). There are preventative steps you can take, for further information see Chapter Eleven.

Dairy breeds

In addition to diet, the breed you choose and the inherent milking ability of the individual animal will have a strong influence on production levels. UK breeds are listed below, in North America Milking Devons and Canadiennes should be on your list, as well as some of the breeds listed which are also present in the US (see Chapter Three).

Included is the average annual milk output of these dairy breeds (where figures are available), expressed in litres per full lactation, which for the cow

Canadienne. (Photo, P.A.B Photographie inc).

means a period of 305 days. This should only be used as a guide since individual animals vary. The figures relate to unselected, traditional type animals kept on a low input, non-intensive regime. Figures in brackets relate to selected animals measured on higher input commercial systems

Small Breeds

Dexter

Average milk yield per lactation: 2295 – 2500 litres, 3.97 per cent butterfat (BF)

The Dexter is one of the most popular dual-purpose breeds for the smallholder. They are relatively low yielding, are very efficient and any surplus stock make high quality beef.

In the UK, the breed has made a comeback from being endangered partly because of these useful characteristics. Dexters come in different sizes, the smallest type may not be the easiest to milk as their udders are very close to the ground, the medium-sized animal may better for milking purposes. Ask advice from Dexter breeders.

Kerry

3250 litres, with up to 4.00 per cent BF

Although relatively small, the Kerry is larger than the Dexter (the Dexter is believed to descend from the Kerry). This breed is on the RBST priority list of endangered breeds and is in great need of support. It is particularly important in the case of the Kerry to conserve dairy potential within the breed.

The Channel Islands Breeds

Guernsey

3660 litres, at 4.5 per cent BF (5095 litres at 4.68 per cent BF)

You would be well advised to seek the more traditional type Guernsey, as the larger North American-influenced animals are designed primarily for intensive production systems. As with the Jersey they produce wonderful creamy milk and have strong selling points.

Jersey

3458 litres, at 4.9 per cent BF (4771 litres, at 5.36 per cent BF)

Larger than the traditional Guernsey, the Jersey produces milk with a higher butterfat content than most other breeds, and in relatively large quantities. If you want distinctive butter, cream and perhaps ice cream, they are one of the best breeds to choose and offer a number of unique selling points to help with marketing your product.

Shetland 2 – 7 gallons per day (no other data available at present)

The Shetland has an excellent reputation as a house cow. They were the mainstays of the island crofting families, are very hardy, will do well on poor forage and generally have an excellent milking temperament. The Shetland is now very rare and needs considerable support. The breed also produces excellent meat.

Larger breeds:

Gloucester

By managing and feeding for production, a cow of a good dairy strain can average 8850 lbs per cow, with average butterfat of 3.9%, protein 3.3%, lactose 4.6%

The Gloucester is a magnificent rare breed, famous for its original use in producing Double Gloucester cheese (NB commercially produced Double Gloucester cheese is made from 'commercial' milk). The milk contains small fat globules and tends to have a high protein content, which is good for cheese making. The breed is now used mainly for beef, so you will have to specify that you are looking for 'milky' types, they are still available, and again, you will be helping to maintain this attribute within the breed.

Irish Moiled

4000 – 4500 litres (no current data available on BF content but a high performing cow is recorded as having produced 8818 KG at 4.3 per cent BF; (L. Alderson, *The Chance To Survive*, Pilkington Press, 1994, p. 101.)

A beautiful dual-purpose breed, which has however been selected more towards beef. Milky strains do survive, so be sure to specify your requirements.

Red Poll

3880 at 3.73 per cent BF (4531 litres at 4.02 per cent BF)

Classified as a rare dual-purpose breed, most Red Polls are kept for beef, but dairy herds still exist. Excellent cattle for rearing several calves (multi-suckling). Try to avoid the 'improved' types which have been crossed with other breeds, if you wish to help with the conservation of this traditional breed. Strongly recommended if you want a long-lived predominantly dairy enterprise with a good beef sideline

Dairy Shorthorn

4250 litres at 3.6 per cent BF (5695 litres at 3.85 per cent BF)

The world-famous Dairy Shorthorn is strictly speaking a dual-purpose breed. Although not rare, Shorthorns have lost out to industrialisation. Some committed producers have remained loyal to the breed. Try to find the traditional type if you want to contribute to their survival, also the economics of this type are more suited to the smaller enterprise.

Ayrshire

4384 litres at 3.86 per cent BF (5879 litres at 4.01 per cent BF)

This breed also has a long commercial history, but they have also been developed to compete with modern Holsteins and so it is difficult though not impossible to find traditional examples.

British Friesian

4825 litres at 3.68 per cent BF

The pure British Friesian, free from modern Holstein influence, is an excellent dairy cow, known for high productivity, longevity and good conformation Thankfully they are enjoying a resurgence in popularity. They are excellent for crossing with beef bulls such as the traditional Hereford to produce meat animals.

The pedigree house cow

It is not imperative to have a pure pedigree house cow, but there are a number of advantages in so doing. Traditional breeds are generally hardier, longer lived, do not require large amounts of bought-in food, are good reproducers and have better conformation (meat yield). Purebred offspring will probably have a higher market value both as livestock and for meat.

In the case of some breeds such as the Red Poll you can, through AI, use bulls which carry proven milking potential in order to breed your own 'milky' replacement heifers.

Last but by no means least, some of these breeds are endangered and do need considerable support. An added bonus to keeping traditional cattle is that others who keep your breed will be only too pleased to advise and encourage you.

Wick Court cheese maturing, including authentic Double Gloucester made using Gloucester milk.

Chapter Eight
Breeding

Breeding your own cattle is your opportunity to not only create a 'cash crop' of calves but also to breed future generations for your herd.

There are many traits or qualities that you can influence in your herd through careful selection of bulls, or for AI, semen from bulls with proven characteristics or potential. If you are looking to improve milk yield you can use a bull (or semen from a bull) likely to pass on improved milking potential to his offspring. For meat you may wish to influence quality, quantity, distribution and growth rates or a combination thereof. The other area for consideration when breeding selectively is the 'soundness' of the future generations. You may have a cow which you value very highly but wouldn't it be nice if, for instance, her offspring had harder wearing hooves than their mother to cope with your rough ground; a slightly better, less pendulous udder or better teat placement for milking. With a small herd, attention to detail, some time spent studying the subject in books and learning from other experienced breeders, it is possible to achieve stunning results.

As with all stockmanship, skills increase with experience, while there is plenty of science involved, there is also a degree of 'art'. Your most important tool will be observation combined with some basic principles. It can be easier to breed from dairy animals because they tend to express their breeding condition more obviously. Beef animals can be more elusive in showing signs of their readiness to breed.

There are two methods of breeding

1. natural service by a bull
2. artificial insemination by the expert use of frozen semen

Before going into the details of how to effect a successful mating by either method, we should focus on the breeding females.

Heifers

The first important decision to take is how old should the 'maiden' female be before you attempt to breed from her? In large-scale commercial herds there is considerable time pressure, however although a small-scale enterprise cannot afford to waste time or money, it is possible to take a longer term view. If an animal is well grown, it may be reasonable to try and get her in calf at 18 months of age, but others in the group may not be so well developed. Some breeders wait as a matter of course until an animal is 28 months old. In the case of Traditional Herefords the ideal age could be between 22 and 24 months old. Breed will have an influence, talk about your animals with others in your breed to help you decide.

There is fierce debate over whether breeding young has any adverse long-term effect on the female, for example does it check their ability to reach their full growth potential? Breeding young can, in my experience, check development but these animals do generally make up for lost time. One of the most important things to consider in terms of the age of the breeding female is that if they are too small they may be more prone to calving difficulties.

Fertility

It can be easy to assume that an animal is either fertile or not, however this is not the case. Fertility levels in your animals can be improved, particularly through the implementation of careful feeding regimes or the improvement of pasture.

The first thing to do if your herd has fertility problems is to check your pasture for any vitamin or mineral deficiencies and take the necessary corrective measures. Some mineral licks are formulated for higher fertility or for specific local conditions. The quality of your pasture may be improved in the long term through the careful addition of clover (see above for the health risks associated with clover) or the use of special herbal seed mixes which should improve the natural fertility of your animals. Another good way to improve fertility is to spread calcified seaweed on your pasture.

It should be pointed out however that your animals are individuals. No matter how much has been done to improve condition or fertility levels there is still a chance that one of your animals is infertile. For example, a heifer which is twin to a bull calf is known as a freemartin and the majority of these are born infertile, as they receive excessive levels of male hormones whilst in the womb. It could be that a bought-in heifer with fertility problems is in fact a freemartin. There are a number of causes of infertility, both congenital/physical and environmental, therefore discussion with your vet will help you reach the right conclusion with a difficult breeder.

Nutrition

The time of year and your management system will have a direct bearing on the nutritional status of your cattle, their fertility level and their ability to calve naturally without assistance.

If they have been housed all winter on a plentiful diet they will be in very good condition but be careful not to overfeed. If they have been outwintered and the winter has been a long, wet one, the animals may be in fairly poor condition. Traditional breeds and types tend to make a remarkable comeback once spring arrives. The ones in poorer condition tend to be better breeders; very fat cattle can be difficult to get in calf. Generally speaking, it is best for breeding animals to be on a rising plane of nutrition whereby the level and quality of their feedstuff is improved gradually. This will be a natural progression in the spring and summer, but if you are looking to optimise their breeding condition at other times of the year, careful supplementary feeding will be necessary.

Oestrus (Heat)

Oestrus is the technical term for your animals' breeding cycle. On average, cattle come into season every 21 days until in calf, and it is this which is called the cycle. Most cows cycle for approximately two days, some have regular cycles, others less so.

It is most important that you make notes of any behaviour which suggests the animal is approaching or is in season. It is also a good idea to keep a small notebook and pencil with you whilst you are going about your everyday work to note down any observations.

Once you have established that an animal is on heat, she can be checked once every three weeks to see if she is on heat again or is in calf, you will then be able to calculate an approximate calving date.

Signs of Oestrus

There are some obvious behavioural and physiological signs to look for

- sparring with other animals
- bellowing (to call the attention of a bull)
- mounting or riding other females and showing more affection than usual
- a raised tail and a swollen vulva
- a 'bulling' string or stream, this is a clear mucous discharge from the vulva

Sign of oestrus is a raised tail and a swollen vulva.

It should be noted that most of the obvious signs of oestrus occur in the late evening often around 10 pm – this fact is generally accepted but not particularly widely known.

Other methods of detecting oestrus (heat)

Self-adhesive 'indicators' are available which can be attached to the back of a cow so that you can see if she has been ridden during your absence, but these are not foolproof, she may have rubbed her back on a branch!

There is also a paste, like the 'raddle' marker (colour marker) used on rams for sheep, to put on your bull, but again this is not foolproof, he may have been keener than her and there may not have been a successful mating.

Lastly, you can send a milk sample to test for the presence of the hormone prostaglandin. This is a popular method with large dairy herds, but it may be less practical for smallholders.

Oestrus can be obvious with dairy types, but some it is more difficult to detect with beef animals. In addition dominant females are less likely to be 'ridden' by their subordinates, and since riding is one of the most visible signs of oestrus you may find it harder to tell when the matriarch is 'bulling'. Some individuals have very short 'seasons' (these are sometimes known as 'shy breeders') others quite long. Some animal's cycle like clockwork, others may be more irregular and may benefit from being examined by your vet, who can sometimes solve simple problems by nothing more than careful manipulation. You will get to know your individual animal's tendencies over time.

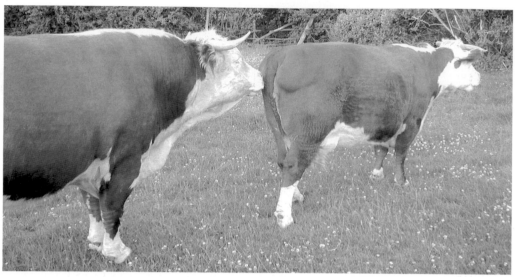

Nature knows best.

A cow will be on heat for about 35 hours. If you are using a bull, he and the female will take the guesswork out of the process by making several attempts to time things right, but with artificial insemination you will need to be organised.

Artificial Insemination (AI)

The optimum time for insemination is towards the end of the heat period, (10 – 20hrs after the heat begins). If you see your cow bulling late afternoon she should be inseminated the following morning and conversely if seen in the morning she should be inseminated in the afternoon. To increase your chances it could be a good idea to have her inseminated twice, the extra expense is far outweighed by the advantages of doing this. (Obviously this is not a practical solution if you are using rare semen in limited supply). The period for which the cow is fertile enough to conceive is only a few hours, and for heifers the period may be as short as two hours, therefore the timing of the insemination is critical.

On-farm gene bank, using liquid nitrogen flask.

For artificial insemination to be a swift and painless process it must be carried out by a professional, training courses are available but it is unlikely, even after a 3 - day course, that with your small number of animals you could ever gain enough regular experience to be a good AI operative.

There are a number of companies who carry out this work every day. You will need to arrange with your nearest approved AI centre to have the semen you purchased transferred to them, they will store it for nominal charge until it is needed. A phone call to the centre in the early morning should see someone arrive within a few hours to carry out the procedure very cost effectively (£12 plus VAT is a typical charge per animal at the time of writing in the UK). Contact your local DEFRA or agriculture office to find your nearest approved AI Centre. Make these transfer and storage arrangements at least four weeks in advance to avoid disappointment.

Single doses of semen, known as straws, can range in price from as little as £5 each to many hundreds for high value commercial breeds.

Liquid nitrogen flask showing straws containing frozen semen.

The advantages of artificial insemination

The cost of transfer and storage of the straws followed by the insemination will be far less than keeping your own bull for a few females. The other great advantage of AI is that for very little outlay depending on breed, your herd can enjoy the considerable benefit of having the genetic influence of a bull which may be or have been worth many thousands of pounds and which would otherwise have been out of reach. In the case of rare breeds you can select bulls which will help minimise the potentially negative effects of inbreeding.

It is important to talk to the breed societies and other breeders when considering the choice of bull. For example, certain bulls will be best avoided for inseminating heifers if they are known for producing very large calves or you may discover that a particular bull carries a defect or potential for mis-marking.

The benefits of AI are far reaching, however this needs to be balanced with the extra trouble AI may be compared to natural service.

Controlled breeding or synchronisation

Synchronisation is a modern technique which significantly increases the probability of AI being successful, particularly important for conservation purposes. It is not the most natural approach and may not be approved for use in organic systems although I have obtained a dispensation from an Organic Certification body for use for conservation breeding of rare animals within an organic enterprise. It can be a useful tool however and in some cases it may mean the difference between an animal breeding or being declared barren and so having to be culled. Your vet will implant a removable hormone implant, the most common of which is called Prid™, into the cow's vagina. The procedure is painless and simple and the implant is removed easily at a predetermined time when the cow is to be inseminated making management much easier since calving can be planned. The procedure is not infallible but you will certainly have increased the odds of success with your shy breeders, a conservative estimated success rate would be around fifty per cent. Feeding regime pre-treatment is also important, your vet will advise, and this could also influence success rate.

Besotted!

The Bull

Some animals do not respond to AI as readily as they will to a bull for not only does the male play the essential role at the time of conception, but his behaviour and attention seems to be necessary in some cases to bring the cow into season.

If you do have or are able to borrow a bull you will need either a relatively distant field or have secure accommodation for heifers which are too young to get in calf. The maximum age to leave heifer calves near a bull is eight months, they have been known to get 'caught' at six months but this is unusual. Do not be complacent about this, the chances of a difficult or even fatal calving in a heifer which is too young can be very high.

The bull will also need separate, secure accommodation at times, but make sure it is of a good standard. Monitor the condition of his feet and attend to any problems without delay. Health and safety should be kept in mind at all times when dealing with a bull. You are strongly recommended to acquire supervised experience both in fairness to the bull and yourself, you will also need to acquaint yourself with the regulations for keeping bulls. Bulls cannot be kept in fields which have public footpaths running through them. It is also a good idea to display a sign stating 'Bull loose in field'. In the UK contact the Health and Safety Executive and DEFRA for the appropriate information or current legislation.

Fencing needs to be strong to ensure bulls cannot escape, if in doubt supplement the weak spots with electric fencing, especially if your neighbour's females are nearby.

An alternative to keeping your own bull is for you to hire a bull for a short period. However hired bulls are often not sent back in good condition, which discourages such cooperation other than with commercial hirers, but by paying attention to the condition of the bull while you have him you could ensure this was not the case. In the UK expect to pay a fee of £15 – £20 per cow to be served, plus transport to and from your place. There is a risk of disease transfer, but you can assess the risks, both ways, with the owner and your vet.

The final option would be to send your animals to a bull. Bull owners in the UK are less inclined to let cows travel to bulls since the recent outbreak of Foot and Mouth, but confidence is beginning to return and movement restrictions have been adjusted.

Care must be taken in not putting too large a bull on too small a female. It is not unknown for the heifer to be terminally damaged by having her legs splayed under the weight. Small females will need a lighter bull than mature cows.

Remember that the bull will have a considerable influence on your herd, so it is a good idea to discuss your choices and options with more experienced breeders if possible.

Timing of calving

Breeders in the commercial sector aim for their animals to produce one calf per year. Cows are pregnant for around nine months and usually they are not put to the bull or AI before two months after calving. There are exceptions however, for example if a cows loses a calf, but she is unlikely to hold to the service of a bull. Smaller enterprises may not be as demanding on their animals' productivity as the commercial breeders. However, it is not desirable to run cows empty for long periods as they can be harder to get back in-calf and may gain too much condition – as mentioned previously, fat animals are less successful breeders. They lay down lots of internal fat keeping all the goodness for themselves rather than sharing it with a calf. It is worth remembering that in the wild, cattle would be in-calf consistently.

Commercially, breeders adopt different calving patterns to suit their production and management needs and demands. Some prefer autumn calving others spring calving, there are advantages and disadvantages to each option.

A great result.

A textbook birth. Note the string from the vulva – do not be tempted to remove it , let nature run its course.

Spring calving means the youngstock will benefit from a whole summer at good grass and the cows' condition is likely to be fitter and more controllable. Autumn calving means that you'll have cows which may be overweight after a summer at grass, with the problems during calving that may occur as a result of too much internal fat. Also you will have vulnerable young animals having to contend with the worst weather of the year.

Autumn calving is perhaps of greatest advantage to commercial breeders. Calves housed in the winter will not grow very fast, but will grow extremely quickly as soon as they are put out to grass in the spring and can therefore be 'finished' quickly. The quality of the beef may not be as good as it is with slower systems, but it could be useful way to boost cashflow. The disadvantage is that there is a greater risk of infection in housed animals during the winter.

On a small scale timing may not be so critical. If you lose a calf, why wait for a particular time of year, particularly if you are breeding for conservation purposes? Your own circumstances and location will be the deciding factors for your plans.

Pregnancy diagnosis (PD)

Even the most experienced breeders are often unable to be certain whether or not an animal is in calf, at least until the later stages of pregnancy. Your vet can carry out a rectal examination, to determine the situation, some with a high degree of accuracy. It is also possible during this procedure for the vet to detect any physical abnormalities which may be restricting the animal's ability to become pregnant if one of them turns out not to be in-calf. Talk to your vet about the best times to do this. Your vet may also have ultrasound scanning equipment which is very accurate.

Chapter Nine
Calving

Calving is one of the most exciting events of the agricultural year. Experienced cows will know what to do better than you do. If you have heifers in-calf for the first time they will only be following their instincts, so it is wise to pay closer attention to them.

You will not be able to watch the cows 24 hours a day, and they may well give birth when you are not around, but there are various steps you can take to increase the odds of all going well. The key to safe calving is observation and seeking professional help in a timely manner if you have any concerns. Your judgement will improve with experience, however it is worth remembering that even the most experienced breeders do lose cows and calves.

Calving dates

When you purchased your animals the breeder will probably have given you a good idea of when they were 'bulled', if natural service was used and if they were artificially inseminated, there should be close to an exact date. The calving date however cannot be predicted completely accurately, as with other animals the gestation period can vary, but at least you will have a 'window' in your diary for when you must be on high alert and preparedness. The average gestation period for cattle is 283 – 284 days, the longest recorded is 439 days, the shortest 200 days (calves born alive), most cattle are around the average.

Calving dates and stage of pregnancy can be established using two methods. The animals may have been 'pregnancy diagnosed' (or PD'd) by the breeder's vet to ensure they were in calf before being sold as such. Some vets are extremely good at predicting the stage of pregnancy but others are less so. The second option is to get your vet to carry out an ultrasound scan. Nothing is worse than watching an animal get bigger and bigger, only to discover it is not in-calf and has simply got so large as a result of having too much food! The decision to check is yours. If the people you bought from are confident that they are in calf (they may even have guaranteed this) then it may be preferable not to put the animals through the stress of a rectal examination.

Healthy newborn calf. (Photo courtsey of Paul Dowlman.)

Nutrition and condition

It is all too easy to overfeed the pregnant animal. You will be inclined to think that she needs all the food she can get but this is not so. If you overfeed her you could be increasing the possibility of calving problems. Too much feed in the later stages will be likely to produce a very large calf, a cow may cope, but a heifer could have problems.

Assuming that any mineral deficiencies on your land have been remedied and the animal has access to a balanced mineral and energy supplement, then her requirements will be very basic. The time of year will play a part in deciding the type of mineral lick you provide, be advised by your merchant or vet.

The cow must be fit not fat, if she has laid down lots of internal fat this can also cause difficulties at calving as the actual space within her can be drastically reduced and this will limit the space for growth and parturition (birth). You will need to keep in-calf cattle on a different feed regime to that of dry cows or cattle you are rearing/finishing for meat. They should not be starved, but bear in mind that in their original wild state they would not have been fat when reproducing.

The time of year will also influence exactly how you control their condition. If it is winter and they are housed, the in-calvers may need to be separated from the rest of the herd. A basic hay ration plus access to molassed minerals will keep them just right. In the later stages of pregnancy the animal will not be able to consume as much dry matter due to reduced internal capacity, that is when judicious use of concentrate/hard feed can be really helpful as the demands on her metabolism increase. If feeding good silage remember that this is a much more nutritious feed than straight hay. Fellow breeders will be able to help you draw up a detailed feed regime for in-calf animals for each breed and at every time of year.

One final point to note if your cows are housed. Due to milder winters in the UK over recent years infestations of external parasites, sucking and biting lice, have appeared to become more severe. It is normal for animals to have lice, but if they are badly infested it will affect their overall condition and add to their burden. If it is spring or summer and they are out on rich grass, they may not find these infestations a problem. However, if they are very dirty at the back end it is possible that the grass is too rich and passing through them so quickly that they are not actually making best use of it nutritionally. In such cases, if you have some straw or average hay lying around, let them have access to it from time to time, you may be surprised how eagerly they consume it and it will help to dry/bind them up a little. You may also follow a 'ranch' approach and just let nature take its course.

Observation (again)

Once again observation will be your most important aid. With beef-type cattle, the visible signs of approaching calving can be very subtle up to the final stages, with dairy types signs can be easier to detect.

Look for the following:

- The earliest signs are a softening and lengthening of the vulva, quite often accompanied by a clear jelly-like discharge. If this discharge is coloured it may indicate a problem and you should seek a professional opinion.

- The next sign is increasing prominence of the pin (hip) bones. This is more difficult to notice with a fat animal.

- Check the udder by hand for signs of swelling if she is accustomed to you inspecting her. With beef breeds they may literally 'bag up' at the last minute, in other words the udder will swell and appear to be full rather than empty and floppy as it is during the cows 'dry' period (i.e. when she is not feeding a calf or being milked) but you can feel for more subtle changes. With dairy types the swelling of the udder is usually easy to see.

- The clearest sign is when the cow goes on her own to a quiet corner of the field. You will need to keep a watchful eye on her, the birth could be imminent but it is just as likely to be many hours away. You just do not know, so be 'on duty' as long as necessary.

Natural Calving

Everyone hopes for a perfect calving without the need for intervention. If your animals are quiet, not stressed too often and are not overfed then the likelihood of encountering calving difficulties such as malpresentation (see below) should be minimal, although first-time calvers may need some experienced assistance. Traditional breeds tend to calve more easily than highly developed commercial breeds.

When to intervene

Depending on the time of year and the weather, your first decision is whether to move the cow to a clean, dry stall if she looks as though she's ready. Her confinement means that she is unable to escape if you do need to help. However assisted births are possible in the field.

It is generally preferable to let nature takes its course but if the protective fluid-filled membrane bag ruptures, her 'waters will have broken'. If there has been no further progress since, or if a nose and one or two hooves are showing but again there is no further progress, it is prudent to consider gentle assistance.

Malpresentation

Normal presentation at calving is for the calf to come head and front feet first, anything else is a malpresentation and you should seek professional help until you are a lot more experienced. You will find excellent reference books in the Further Reading section.

Internal Inspection

A simple internal inspection can be carried out to establish the precise nature of the malpresentation without great risk to the cow, as long as simple precautions are routinely taken.

1. Make sure your hands and nails are thoroughly clean. Keep nails short.
2. Use an antiseptic lubricant gel, to help minimise infection and damage to the inside of the cow.
3. When manipulating the limbs and/or head of the calf, take care not to cause internal damage to the cow, use your hands to protect the delicate internal tissue of the cow.

Intervention

Calving Aids

Calving ropes which are available commercially and are made of plaited nylon, they can be washed and reused but must be stored hygienically

Antiseptic lubricant gel

Iodine or antibiotic spray (from your vet) for navel dressing

Weak calf reviver (not everybody keeps this in stock as it can be acquired easily from your merchant)

Antibiotic (injectable, plus needles and syringes from your vet, who will advise on quantities and gauge)

Dry cotton towel (s)

Warm water, soap and scrubbing brush for you and/or the vet

Calcium/magnesium solution available from your merchant or vet and 50 ml syringe, in case of milk fever (parturient hypocalcaemia see page 91) which can occur after calving and particularly at certain times of year. Some breeds are more susceptible, discuss with vet.

Calving jack. This should only be used when absolutely necessary and with skilled help. Your vet will know when this is necessary, or your more experienced neighbour if a vet is not available. The larger enterprise should consider having one as part of the emergency on-farm kit.

Assisted delivery

1. If calving ropes are needed, they must be attached above the calf's ankle joint to prevent injury. You can also attach a bar or stout stick to the ropes to give yourself uniform leverage. You will need to use ropes if you have to intervene in a determined way, obviously the calf is slimy and wet so it is difficult to get a firm grip by hand. The ropes can have a few knots added to improve the ability to grip them as they can get slimy too.

2. If you have decided to fully assist the delivery, take your time, and always pull down rather than straight. Pull with the cow's contractions, not against her. Note that the cow may remain standing for quite some time, but she may well decide to lie down, even on one side, do not be put off, you may even find helping easier in this position.

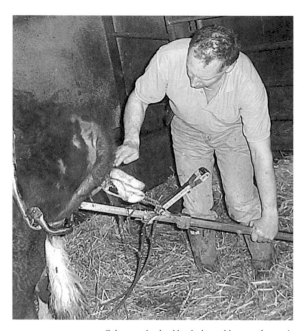

Calving jacks should only be used by a professional.

3. If you have had to intervene at calving, or have carried out an internal examination, a precautionary antibiotic injection should be administered to prevent infection.

Further study on calving is recommended, please refer to the Further Reading section.

The newborn calf

If the new arrival is being carefully attended to by its mother all you should do is watch for a while to see that all is OK. The process of the cow licking the calf is part of the natural way to encourage it to get to its feet as quickly as possible. It is a good idea to see that the nose and mouth are free of any detritus so the risk of suffocation is reduced. A first-time calver may neglect to see to this. Do not worry if the calf is shivering, the licking process, despite being wet, warms the calf up. Always satisfy yourself that the calf has taken its first feed before you begin to lower your guard. This can take quite a while, so don't panic, it should normally happen naturally within half an hour. If it has been a long and/or difficult birth the calf may be quite tired. In such a case if you feel matters are not progressing, you can milk out some colostrum and give it to the calf by bottle, or stomach tube. You will need to be shown how to use one by an experienced person. The results of this type of intervention are usually very rapid.

Colostrum is a very rich form of milk which contains a variety of antibodies essential to giving the calf immunity to disease. It is produced naturally by the mother after birth. The calf must have colostrum within the first 24 hours to stand the best chance of growing into a

A cow cleaning her newborn calf. Note that the nose and mouth are clear, so there is no need to intervene.

healthy animal. In case the worst ever happens and you lose a cow, it is not a bad idea to milk out some colostrum at some point and freeze it as an emergency stop gap measure. It can be stored in a domestic freezer for up to a few months. There are commercial colostrum replacers which you mix as required.

You will know that the calf is ticking over nicely and feeding off its mother successfully when it produces its first poo! This soft sticky smelly substance is known as meconium. Note that if the meconium is particularly foul smelling, very runny and pale it may suggest some kind of infection, seek advice without delay.

Heifers

An experienced cow is very unlikely to reject her new-born calf, but with heifers there can occasionally be problems after calving and it is therefore advisable to keep any intervention to an absolute minimum. In the case of assisted births it is important to present the heifer with the calf as soon as possible. Confining the animal with her calf is another option if she does not appear to be attending to it properly.

In cases of rejection you may need to remove any membrane covering the calf's mouth and nose to prevent suffocation as soon as it is born. If the calf looks lifeless take a towel and rub it briskly to help to 'get it going'. If you still aren't getting a response, lift the calf up by the hind legs to help any fluid on the lungs drain, you can even swing it around vigorously to aid this process. Another common practice is to slide a stem of straw up the nostrils to help clear any mucous, this shows good results.

Last but not least, your vet or merchant can provide you with weak calf reviver. If, despite your and the cow's best efforts the calf is alive but non responsive it may be worth a try

Navel ill

Always allow the umbilical cord to separate naturally. NEVER try and remove any remaining cord from the cow as this can cause real problems. Confer with your vet if she does not 'cleanse' within a certain period as he can give her an injection to help.

Once separated, as a precautionary measure always dip the calf's umbilical cord in neat iodine to prevent infection entering the cord and causing navel ill.

Safety issues

If your cow and calf are housed, do NOT be tempted to cover the calf with straw to keep it warm however cold it is, there is a good chance the cow may crush it inadvertently, find some other way to reduce draughts. It can pay to leave a light on for the first night at least for similar reasons.

Make sure that your housing or fields are 'calf safe'. There should be no bits of loose baler twine for it to chew and choke on, and any gaps in fencing should be blocked and potential hazards like loose bits of barbed wire should be removed.

Post-calving care of the cow

Ensure that the cow ' cleanses' properly once the calf has been born, that is to say that she expels the afterbirth. This can happen quite quickly after calving, within an hour or two, or it may take a few days to be completed, particularly if it has been a difficult calving. It is not unusual in cases of difficult calvings for part of the afterbirth to protrude for some time. Do NOT under any circumstances be tempted to try and remove it, however horrible and smelly it might be, you can only make matters worse by taking such steps, the cow should be allowed to cleanse herself. If you are worried, seek professional advice.

The cow will usually try to consume the afterbirth after expulsion has taken place. Some believe this to be an important and necessary event which acts as a trigger for milk production and that the afterbirth contains vital nutrients. However it is more likely to be due to instinctive behaviour based on the cow's need to reduce the risk of attracting predators to her vulnerable calf.

Warning

It is not unknown for the cow to choke on the afterbirth, and in some cases this can be fatal, and can happen in a matter of minutes. If you are concerned it can be a good idea to try to remove the afterbirth immediately after expulsion to minimise the risk.

Problems and hazards

You will need to watch the cow for any signs of discomfort over the coming weeks even if calving and cleansing have been normal. It is normal for her to have a bloody discharge for a few days whilst she recovers from the rigours of birth, but if she starts to smell bad, arches her back, and holds her tail up a lot, she could have an infection. Take her temperature. Normal temperature is 38.9°C (102 °F). If her temperature is above or below this for a significant time you should call the vet.

Finally, if you have had the misfortune either to lose the calf or the cow as well you will need to be philosophical. Disasters happen to even the most experienced cattle people.

Chapter Ten

Calf Rearing and Meat Production

Perfection. North Devon Beef in an attractive display.
(Photo courtesy of Ann Petch of Heel Farm Meats, Devon.)

One of the harsh realities of livestock production is the fact that not all the animals produced are either of high enough quality or in sufficient demand to become part of the next generation of breeding animals. In addition, fewer bulls than females are required, particularly as a consequence of the increase in artificial insemination.

As a smallholder, even if you do not have a house cow or breeding herd, you may like to consider rearing two or more calves for your own and/or neighbours' consumption. You may even choose to rear small batches of calves regularly to supplement your income, a lot will depend on your circumstances and facilities.

Types of meat

Natural veal

I use the term 'natural' advisedly as in Europe in particular intensive veal production is more or less an industrial process. As such it is if no interest or value to readers of this book.

What you can consider is the production of veal by non-intensive means and this is a product which is far superior to factory-farmed veal.

A term which is popular for this type of meat is 'grey veal' – it is not white or pink as meat from animals which were housed from start to finish and is from young animals which have done some grazing. It could also be called 'young beef'.

It is possible to use calves known as 'poor doers', which we all have from time to time, and which will not make prime meat or breeding cattle.

Once you have tasted this type of veal, it is easy to see just how special it is. As a bonus, the animal has lead a nice natural, albeit short, life. You won't have to worry too much about conformation either, as long as the animal has got a reasonable amount of meat on it you should make money. You will be able to charge a premium price for this meat as it is a delicacy, and demand should easily exceed supply, with plenty of repeat orders.

Store beef

The longer-term approach to calf rearing for meat is to produce high quality 'store' beef animals. Store animals can be sold on to 'finishers' as soon as they are weaned at six months old but more often they are sold at twelve months plus. You will need to keep an eye on your profit margin – the cost of the calf, plus all your inputs including time and any losses, will mean there is little room for error.

Finished beef

If you are able to produce a 'finished' animal then your margins should be a little less tight.

'Finishing' means that you have taken your beef animal all the way from being a calf to an animal which is fit and suitable to slaughter. If you sell this animal at a market, the dealer or buyer will get his margin, however if you sell direct to a butcher you will cut out the dealer's margin. To get the full potential you can even sell the processed animal direct and get all the money!

See Chapter Thirteen for a detailed look at how to market your beef.

Meat production

Buying-in calves

It can be an excellent idea to buy in weaned calves from producers of rare and traditional breeds. Contact local breeders who may have surplus stock.

Note: with the traditional breeds and crosses in particular, the heifers you rear do have another potential market as suckler cows, either for you or your neighbours.

There are a number of advantages to buying stock directly from farms in preference to markets:

1. risk of disease transfer is minimised;
2. it is less stressful for the animals, rather than being taken to a market where they will be amongst lots of other stressed animals, put through a sale ring and then delivered to your place, they just make one journey from farm to farm;
3.   you can see the source of your stock, the type of operation it has come from, and what the other animals in that herd look like;
4. you will get to know the producer who may be able to give you advice as well as sell to you.

Other types of calf

In addition to calves from rare or traditional breeds there are a number of other alternatives to consider when planning meat production.

Traditionally, Channel Island (Jersey/Guernsey) bull calves have been regarded as having little or no value on the market. However they can be reared for meat with surprisingly good results, for beef or natural veal as above. It is worth remembering that to a specialist producer like you, conformation is of less importance than it is to intensive producers. Contact Jersey or Guernsey herds in your area to see if they will sell you a calf or two.

If you have not chosen pure breeds it may still be a good idea to look at traditional types or crosses, such as Dairy Shorthorn crosses or Aberdeen Angus crosses although they may be more difficult to track down. These are a better economic option than modern continental crosses (discussed below) which will require a lot more feeding

On compassionate grounds alone I have been tempted to buy some 'bobby' (just born) calves from intensive dairy units, although on a small scale it is probably best to avoid this type of enterprise.

The final type of calf to consider buying in is a continental-cross calf from breeds such as Limousin, Simmental, Belgian Blue and others. This is probably the easiest option as these calves are more common and are sought after, this is reflected in the higher price you would normally have to pay for them as you will be in competition with the commercial producers. Many of these come from dairy herds, others are bred specifically as beef animals from suckler herds.

Limousin crosses do have a reputation for being more difficult to handle than some of the other crosses and are therefore not recommended for the novice.

The continental animals produce a lot of meat which can fetc.h a good price, however it should be remembered that the calves do cost more to buy-in and they require more food than the traditional types. Margins and costs will have to be looked at closely if this option is followed. Bear in mind that the meat from traditional types could be of higher quality if finished and slaughtered properly.

Whilst margins for continental crosses are tight, these animals tend to sell most readily at market, but it may be better to avoid that treadmill if you can.

Interestingly, calves produced by one of our Traditional Hereford bulls, who was used on Longhorn cross British White cows (obscure concoction, see picture on right) on Salisbury Plain, strongly resembled continental Simmental crossbred calves and were strong boned and well put together. They always got best prices at Salisbury market and were in great demand. They cost a lot less to produce than normal continental crosses, yet achieved the best prices against strong competition.

A calf from a traditional Hereford bull crossed with a Longhorn cross British White cow .

Calf rearing

Housing

Your existing or planned facilities will naturally dictate the size of your venture. You should start with a minimum of two calves, they will do better together and increase your chances of success. On this scale, a simple stable-type loose-box or pen will be sufficient for housing, in other words a small 'room' in which they can be confined safely and comfortably, within a larger building if necessary. This needs to be airy but draught free.

If you do not have the luxury of a loose-box and are considering partitioning a fairly exposed part of a larger area such as a barn, you should stop the rain and wind driving in. A few well fastened slatted pallets might do for the basic structure, but you will need to make the pen draught proof. It is very easy to subdivide a barn area with field-type gates, which can be cladded on the outside with bags/polythene or even straw bales to prevent draughts. A false roof suspended above the pen will also help to keep them warm. You could achieve this with some stock netting, poles, straw and feedbags. It is vital to keep the bedding dry and clean.

You will also need to consider how you will house them when they are bigger. The time of year will influence your planning, as they may be able to be out to grass nearly all of the time, which makes management much easier.

Health and hygiene

We will consider general health in Chapter Eleven but it is worth highlighting a few points here. Ideally calves should be with their dams for several days before being separated from them to ensure they have a good start in life. They will recieve the all-important colostrum in the first milk which will provide them with vitamins and antibodies giving them a high level of instant disease resistance. Sadly many calves from dairy units have no such opportunity, so when

Traditional Hereford beef steers, heads in the trough.

purchasing, especially as a novice, try and find out whether they received colostrum for a minimum of 24 hours after birth. Colostrum replacers are available but it is best to give yourself and the calf a safer start with its mother's colostrum.

The animals will be at the greatest risk of illness once they have arrived at your farm as a result of the stress of the journey, unfamiliar surroundings, and any infections or disease present on your holding (that you are almost certainly unaware of) for which they have no antibodies. As long as you have adhered to the housing recommendations mentioned above then you will have minimised environmental risks. Of equal importance is hygiene, all drinking and feeding apparatus must be kept scrupulously clean, however boring a task that is.

Scours

The main killer of calves is scours (diarrhoea). Causes can include bacterial infection, such as E. coli, viruses, such as rotavirus, or simply the stress of being moved, becoming chilled, overfeeding or a rapid change in diet, being fed mouldy food and/or low quality milk replacer. The smell of scours is unmistakable, if you suspect scouring, act immediately. You should dose the animal according to the instructions with an anti calf-scour remedy which you should always keep in stock. Antibiotic injections will be needed too (see below).

General health and welfare

As usual get general advice from your vet before the animals arrive. He may provide you with a vitamin supplement injection, and perhaps antibiotics for quick therapeutic use should a problem arise.

You should also discuss with your vet castration of the male calves to aid management, but this should not be carried out until the calves have settled in well. Note that dairy bulls can be dangerous when they reach maturity if left entire. There are various methods of castration, once you have some experience you will be able to deal with this yourself, but until then discuss the options and timing with your vet. A botched job does not only have severe legal/welfare considerations, you may also discover too late that the operation was not a success when one of your 'steers' (castrated male) got your best pedigree cow in calf when he escaped one day.

Feeding and weaning

It is essential that you do not overfeed the calves and make sure you buy only the highest quality milk replacement powder. Cheap milk formulations may not supply all that the animal requires in terms of nutrition and it is also possible if you do not know the original source or brand, that the milk replacement has not been produced or stored in the most hygienic way. These issues could result in illness for your calves – it is not an area to try and cut corners.

If you have bought the calves from a farm, ask the breeder for advice on how to continue and change over to your feed regime.

Please note that any change in diet should ALWAYS be done as gradually as possible. When you acquire the calves, buy some milk powder from the seller to enable a gradual changeover, even batches of the same brand of milk replacement can vary.

Feed amounts and mixing proportions will be printed on your proprietary feed, mix accurately and always ensure that the feed is given at the same temperature. Calves are usually happy to drink water from a bucket. The water must be kept clean, fresh and be available at all times.

Typical feed regime for a young calf

There are many variables involved in feeding calves, however you will find guidelines on the milk powder bag and these will include suggested feeding frequency. In the early days you may well have to feed them in the middle of the night as well as during the day, but as they grow (quickly) you can spread the feeds out to more convenient times. Remember that the calf in the natural state has the cow 'on tap', so regular small feeds to begin with are better than two or three massive feeds.

A healthy calf will drink all you offer it to bursting point but this is to be avoided. As a guide look at its flanks from each side; if the flanks appear hollow then it needs more, if the hollows have been filled it has had enough, if it bulges either side slightly then you have reached the limits, if it is as round as a barrel, you've overdone it!

Weaning

Calves should be offered good quality hay (ad lib i.e. as and when they want) from the outset. Better that they nibble at that than their bedding, and although they will not gain much from it nutritionally they will be better prepared for weaning, which is the gradual withdrawal of milk from the diet.

From as early as one week old you can offer a special calf feed mixture often called a 'creep' feed, leave it in front of them all the time, they will not eat much at first but by three weeks they will be readily consuming it and they will be prepared to be weaned at around five weeks.

Creep feeding is the term used for a method of enabling calves to acquire their own continuous supply of feed without other members of the herd helping themselves to it. In a housed situation, you may wish for calves who are still with their mothers to have some supplementary feeding. This can easily be organised by erecting a strong pen within the main pen, which has only a narrow access point for the calves to fit through, excluding the larger animals.

If at five weeks the calves are not consuming enough solid feed as described above (as a rough guide up to 0.7kg per day) continue milk feeding and then gradually reduce the amount of milk you are giving.

At weaning you can change (gradually) to a cheaper, lower protein, grower ration. The creep feed is more expensive than a simpler ration formulated for longer term growing on. If you are rearing larger numbers of calves there are various semi – and fully automated feeding systems available, as well as special calf creep-feed dispensers, enquire at your local merchant.

Age at slaughter

Breeds differ considerably in how much feed they need and at what age they will be finished and ready for slaughter. Some breeds will eat enough to reach a saleable standard by 14 months, but the meat will not be of as good a quality as an animal of between 18 and 24 months. Some of the larger or slower maturing breeds will need to be pushed hard to get them 'finished' by the current UK legal limit of 30 months for meat animals. Slower maturing animals however can provide the very best quality meat.

Target 'finished' weights for all breeds are available from the breeders and breed societies.

Chapter Eleven
Health Care and Diseases

Good stock management and the correct feeding regime will minimise health and disease risks in your animals, however you must be aware of potential problems so that you can react to any developing situations swiftly. The timing of your response to a situation is usually the most critical factor when dealing with health and disease.

Cattle can be affected by a number of diseases, but you will have to be very unlucky indeed to be confronted by most of them. However you should get yourself a good general reference book to diseases of cattle so that you can broaden your overall knowledge, as well as have some information to hand at times of concern (see Further Reading).

You should also confer with your vet and ask him to help you prepare a written herd health plan. Having such a plan will help you to qualify for various 'Farm Assured' schemes, should you wish to join one. On a small scale this is a simple exercise so do not be daunted.

The type of health plan you draw up will be dictated by how much land is available, stocking density and whether you are taking the organic 'route' as certain restrictions on the type, application and withdrawal period of medicines, apply to organic systems. If you are going organic you can apply for dispensations to use normally restricted medicines to deal with endemic problems which may affect your holding. Contact your certification body for more details. Your vet will also be able to teach you how to give injections properly, as well as give guidance on administering other treatments.

Local health problems

It is always a good idea to gain as much local knowledge from your neighbours as you can. Your vet will also have specialised local knowledge of potential problems.

There may be hazards specific to your holding. Old cattle sheds may harbour ringworm, a fungal skin disease, and no amount of cleansing may solve the problem completely. Rotavirus which causes scouring (diarrhoea) in calves may also be difficult to remove completely. On occasion, rotavirus may only be controlled through routine use of a vaccine if it is 'endemic' on the farm. This type of routine treatment would not lead to the removal of organic status if it is a recognised problem. (See below for a more detailed discussion.)

Accidents and prevention

If potential hazards on your holding have been minimised, then accidents should be a very rare occurrence.

- Always keep gates to the road closed, just in case the animals escape from the field into your yard.
- Always clear away any netting, barbed wire or sections of baling twine (which cattle and particularly calves seem to love to chew and worse, swallow.
- It may be worthwhile removing broken thorny hawthorn branches and twigs during routine field inspections.
- Block off potential escape routes where an animal can damage itself.
- Rabbit burrows and any other holes in the path of the animals should be filled in. A broken leg can mean the end for that animal. (I have seen cows with false legs, but you need to be quite wealthy!)

First aid

Even on the tidiest holding the occasional accident is unavoidable – the animals cannot be watched 24 hours a day. While some injuries may result in the animal having to be put down, most minor injuries can be dealt with using simple first aid.

You should be able to deal with some simple first aid procedures without calling on the vet or neighbours, at least in the first instance. A limping cow may only have a hawthorn twig stuck between her hooves. A calf with a piece of baling twine dangling from its mouth may not have swallowed it too far so it should be easy enough to remove. Cuts and rips to the skin can be treated with iodine or an antiseptic spray which is part of your first aid kit, but note, it does sting so you won't get any thanks. Any of these simple procedures are easier to deal with if you can coax the patient into a confined handling area as described. If this is not possible, stealth is you best ally; reassuring words, stroking the animal, a little food, followed by speedy action. Having a good relationship with your cattle is a real help in first aid situations.

First aid kit

This is a basic list, your vet may advise the inclusion of additional items. Some items will need to be renewed as they pass their sell-by date, just be glad you didn't need to use them!

  digital thermometer

  scour (diarrhoea) formula (particularly if calf rearing)

  colostrum substitute (optional)

  liquid iodine for wounds, navel dressing, disinfectant

  magnesium and calcium solutions (watch the sell-by dates; you must keep this in stock – by the time you have gone to get some when you need it, it could be too late when you get back)

  16 gauge x 1 inch hypodermic needles

  A quantity of 20ml syringes and a few 50 ml

  bandage and wound dressings

  calving ropes and obstetric gel

  washing soda crystals and or peanut oil (otherwise known as groundnut oil, which, if you cannot get from your agricultural merchant can be bought in litre bottles from retail food outlets)

  glucose powder

  antiseptic spray

  stomach tube kit

  disinfectant

Observation

Observation is the key word in managing the health of your stock. On the busy holding, particularly in winter when the days are short, you may be working flat out just to get all your jobs done, but you must include time to stand at the gate or at the pen preferably twice or more a day to avoid missing any important signals or potential problems, if you do this regularly you will be surprised how quickly you will become attuned to the condition of your stock. Another benefit of this type of daily contact is that the animals respond to their keeper in a wholly positive way, strengthening the bond between man and beast.

In the field, one of the most obvious signs that an animal is not feeling well is that it will stay away from you and the rest of the herd when you call them to you. It should be noted however that young calves may stay away from the gathering if they have only just woken up. If this happens it may be wise to go and check on the calf anyway.

Temperature

The most significant indicator of an animal's health is its temperature, except for obvious problems such as eye infections. If you are concerned about an animal's overall demeanour it is a good idea to take its temperature. Your first aid kit (see page ***) should include a digital thermometer, they are not too expensive and are very easy to read. The reading is taken rectally. The normal body temperature for cattle is 38.5 °C (101.5 °F). Temperature can increase if a female is on heat, but as a general rule if there is any variance from normal temperature i.e. if the temperature is lower or higher than normal for much longer than an hour or two and the animal is displaying other symptoms that are concerning you, contact your vet, giving him all the symptoms and the temperature of the animal in question. Monitor the temperature regularly to see how things are going.

Isolation

Depending on the seriousness and potential infectivity of the ailment, you should always have a space to put your sick animal(s). This provides a warm dry place for the patient, makes it easier to administer treatment and minimises the spread of infection. Although the animal may be stressed by being separated from the rest of the herd, with plenty of attention from you, they could recover more quickly than if they were at the mercy of the elements. If you suspect a contagious disease separate the affected animal sooner rather than later, you can always let them out again if your suspicions proved unfounded.

Homeopathy

Many people now use homeopathy alongside conventional treatments, particularly organic producers. You may be fortunate enough to have a local vet who practices both conventional and alternative disciplines. Whatever your views on this it is vital to remember that an animal should not be made to suffer for the sake of ideological principles – welfare should always come first. I have included details in the Further Reading section for those of you who would like to learn more about homeopathy. I have heard of a number of cases where it has been used to good effect in dairy herds, particularly for the treatment of mastitis.

Ailments and Diseases

Below are most of the more common disorders which can affect cattle, followed by a list of less common diseases. It is recommended that you obtain a good cattle health reference to broaden your knowledge and help with diagnosis.

Hoof care and lameness

If an animal is limping, it should demand your immediate attention. The likelihood is that it may just have a foreign object trapped or embedded in its foot. The damage caused by such objects can lead to a general complaint known as foul-in-the-foot or foot rot, which is caused when a wound becomes infected. If your animals are used to regular hoof inspections, a quick but careful inspection in the field may reveal and solve the problem.

When inspecting hooves, do not go straight to the foot as this will probably make the animal move away quickly. It is much better to greet them, scratch the tailhead for a while and gradually work your way down, being aware that a kick could be forthcoming! If all your attempts at casual inspection fail, monitor the situation for 24 hours and if there is no improvement a thorough inspection should then be carried out initially by you and if necessary your vet, by restraining the animal in a crush. If you do not have a cattle crush, use your gates to make a narrow pen, preferably beneath a beam or branch so that you can use a rope to lift and restrain the leg to enable you to work safely. It can be difficult to see if an object has embedded itself and you may need professional help to diagnose and fully solve the problem.

It may be the long sward rather than a foreign object which has caused soreness between the claws, however it may be wise to give a precautionary injection of penicillin, your vet will advise you. If the lameness is due to an overgrown hoof, again, professional advice and help should be sought. Take care if the animal is heavy in-calf when doing anything which may cause physical stress.

Routine hoof trimming

Routine hoof trimming is not generally necessary for animals which are outside for most of the year. For housed animals, confined to soft-bedded areas, the hooves are bound to overgrow during the winter. Note that some individual animals will be worse than others . Therefore, at turnout in the spring, hoof trimming should be carried out by your vet or a professional hoof trimmer. It is unlikely that you will gain enough experience on your scale to become an expert hoof trimmer, even large commercial herds call in professionals. Leaving overgrown hooves can cause all sorts of strains on the animal so do have them checked; it causes pain to the animals and affects productivity.

Heavily in-calf cows may limp but the cause may not be the foot, rather a nerve may be trapped due to the extra bulk the cow is carrying. The problem should solve itself, but if in doubt call the vet.

Parasites and diseases of the skin

Worming

The presence of internal and external parasites is will add another element of physical stress which will challenge your animals' overall well-being. Cattle cannot cope with these infestations in a domestic situation without some help. If the internal worm burden gets too high, the animal's physical condition will deteriorate. It is worth noting that cattle seem less prone to worm infestation than sheep.

The best way to avoid infestation of parasitic worms is to operate a clean-grazing system whereby the animals are moved to fresh pasture once within a three week period. This breaks the lifecycle of the parasites. In the absence of clean grazing you will need to treat the animals from time to time with wormers, called anthelmintics. Your location and the previous uses of your pasture will affect the frequency and type of wormer you use. It is also important to vary the wormer type, as there are resistant strains of worm. Your worming/parasite control regime will be the mainstay of the herd health programme and you should discuss it with your vet at the outset.

Ringworm

Ringworm is a fungal disease affecting the skin. Humans can catch and also be carriers of ringworm. The ringworm fungus can survive in old buildings, utensils, fence posts and so on seemingly without any time limit and can be impossible to eradicate unless you scrap the building and put a new one up (highly impractical in most cases). You will need to look out for it.

Symptoms: it is typified by animals scratching incessantly, and by round bare patches on the skin on any part of the head or body.

Treatment: several proprietary treatments are available from your farm supply outlet.

Lice

Lice usually affect animals in poor condition, although they do affect healthy stock as well, housed or confined animals are most susceptible. If heavily infested with lice, its coat will be damaged, it will be unhappy and again, its condition will deteriorate.

Treatment: powerful highly effective injectable and pour-on solutions and powders (which are less effective now as certain ingredients are no longer allowed) are available. Some of these products are combined with wormers (anthelmintics) as well. For those who wish to avoid such products, organically approved louse powders are available.

Mange

Management and nutrition can have a strong influence on the prevalence of this disease. It is caused by a mite invisible to the naked eye. It manifests itself by bare patches of coat and flaky skin, the mites burrow under the skin.

Be aware however that not all bare patches automatically mean mange. Cattle, particularly housed ones, do get scruffy coats over the winter, but you will need to be vigilant. Your vet may need to take a skin sample, to be analysed under the microscope, to confirm that it is mange. The vet will then advise the correct treatment.

Eye Disorders

New Forest disease

This is an infection which occurs when an animal has scratched or damaged its eye in some way.

Symptoms: weeping, partially closed eye, becoming cloudy, and gradually more opaque.

Treatment: a number of treatments are available from your vet. New Forest disease is highly contagious, affected animals should be isolated if possible. However, it is often the case that by the time you notice it, it may have started to affect others in the herd. Other eye disorders are often erroneously described as New Forest consult the vet for correct identification of the problem.

Other eye conditions include 'silage eye', conjunctivitis, and chaff-in-the-eye, have similar symptoms and treatments as above. Accurate diagnosis can be made from swab samples to ensure the correct diagnosis and treatment.

Nutritional disorders

Vitamins and minerals

With organic production you are only allowed to provide supplementary vitamins and mineral where there are known deficiencies. Whichever management system you use it is vital that you establish whether or not there are any deficiencies by having your ground tested as referred to in Chapter Five, as well as by gaining valuable local knowledge from your neighbours. The sooner you know about a problem the better and you must discuss the best way to deal with the issue with your vet, as dosage levels are critical in some cases.

Molybdenum

High levels of the chemical element molybdenum inhibits the uptake of copper and this has an adverse effect on both growth rates and fertility. Cattle in these areas need a copper supplement added to their diet or given in the form of injections (this is a particular problem in Warwickshire) but this will not solve the problem. By far the most effective treatment is a combination of minerals administered on a slow release basis. Your vet will be able to advise and help you find sustainable long-term solutions.

Selenium

In the New Forest area selenium levels are severely depleted because this trace element has been absorbed by the trees in the forest over hundreds of years. Low selenium levels also have an adverse effect on fertility.

Cobalt

Cobalt is another trace element which is essential for an animal to thrive, some areas such as Scotland, Northumbria, Devon and north Wales are known to be deficient. Some wormers contain a cobalt supplement.

Salt and mineral licks

It is also well known that salt is important to cattle and it is a good idea to provide access to rock salt for most of the year as well as molassed mineral licks to cover all eventualities particularly if you are out-wintering.

Hypomagnesaemia (magnesium deficiency)

At certain times of the year, particularly autumn and spring, milking cows and cows with calves use up their reserves of magnesium and are then susceptible to hypomagnesaemia, also known as grass sickness, or staggers. You can provide high magnesium licks to help minimise the possibility of this happening particularly if out-wintering cattle.

Note: There is a continuing debate over the possible link with badgers and the spread of bovine tuberculosis. You are advised as a precautionary measure to raise molassed mineral licks to at least 30 inches from the ground to minimise potential disease spread as badgers like them too!

Older cattle are more susceptible to hypomagnesaemia as they do not carry more than four or five days' worth of magnesium reserves. It can also affect younger cattle fed on milk alone as the magnesium content of the milk may be variable and is difficult to monitor. This condition may suddenly appear in outwintered cows particularly after sudden cold spells or in early spring when their reserves are at their lowest.

In the spring when the herd is turned out onto pastures which have been fast growing after fertiliser application, the high nitrogen content of the grazing can inhibit the animal's ability to absorb magnesium. This is also the case if it has rained and the grass has grown quickly.

Symptoms: shivering, staggering, falling down and having a fit, kicking, frothing at the mouth.

Prevention: magnesium levels can be raised through careful use of artificial fertilisers. The pasture can be top-dressed (applied to the surface) with magnesium limestone. If there is a major problem on the holding, magnesium oxide can be added to the feed.

Treatment: in an emergency call the vet who will administer the necessary magnesium.

Hypoglycaemia (also known as Bovine Ketosis/Acetonaemia (reduced blood sugar levels))

Hypoglycaemia may occur when a cow is producing milk at high demand and the level of nutrients she is taking in are insufficient to keep up with the level of production. Swift action is needed.

Symptoms: loss of appetite, constipation, distinctive sweet smell on the breath. Temperature normal. Dung is often coated with slime.

Treatment: you should call the vet as major injections are needed to cure this disorder. While you are waiting for the vet to arrive it may help to drench (dose orally) the animal with glucose solution.

Administering a drench

A plastic cooking oil bottle can be the easiest way to administer a drench. You will need to have the animal restrained (on a halter at least). Stand by her side behind the head, pass one arm over the neck and reach round and lift up the head by the muzzle. This which will result in the mouth being opened. Insert the bottle into the mouth and pour, being careful not to send too much liquid in at once to avoid choking. If the animal is particularly uncooperative, you may need to take firm action, you can force the muzzle up by using the nostrils to gain purchase, but by this stage you will need to complete the task as quickly as possible as she will like it even less than your first attempt. With really difficult animals get an assistant to pour as you concentrate on keeping the mouth open. Using a plastic battle also means you can squeeze it to adjust the flow rate to suit the circumstances. You need to be assertive to dose an animal.

Hypophosphataemia (phosphorous deficiency)

Symptoms: Cows in late pregnancy are most at risk. Mild cases are hard to detect, an inability to get up being the most obvious sign. Appetite may be unimpaired.

Prevention and treatment: establish mineral levels on your holding through thorough testing. Provide palatable mineral supplements to counteract any deficiency. Call the vet immediately.

Hypocalcaemia, also known as Milk Fever (calcium deficiency)

This usually affects older cows within 24 hours of calving.

Despite all your best efforts to avoid it, milk fever can and does strike. What you must do, however is act swiftly in your treatment of it.

Symptoms: loss of appetite, cold ears, shivering, unsteady weaving walking pattern, constipation, heavy breathing, staggering and eventually going down. The animal may then start tossing herself about and the final stage is coma.

Treatment: as soon as you suspect milk fever, inject 448g (16oz) of calcium solution (first aid kit) under the skin, massage the site following injection to disperse the solution. Do this without delay; do not wait for the vet. To prevent pelvic damage, hobble (tie together) the back legs to avoid splaying. If you have discovered the animal after it may have been affected for some time, it may be bloated and you may need to take emergency action (see section on ordinary bloat below).

It is essential that you call the vet if any of your animals show the above symptoms as the condition may not be a simple calcium deficiency.

Note: the calcium solution you have in your first aid kit will cover other deficiencies, such as staggers, as well.

It is essential that you seek professional advice and support. If this condition is a recurring problem on your holding your vet will advise on future herd strategy and the longer term nursing of your patient.

Poisoning

Hopefully, through generations of good husbandry, your holding will be free of plants which are dangerous to livestock, but you will still need to be vigilant, particularly if you are reclaiming long disused pasture, woodland or parkland for use with your stock. Below is a list of the most dangerous plants to avoid:

yew tree

rhododendron

acorns

ragwort (an increasing problem due to reduced control measures nationally. Neglected land you are reclaiming may be rife with it, acquaint yourself with it, remove it on sight, and take it away with you, in dried form it is even more lethal)

laburnum

bracken poisoning can unexpectedly affect stock that has lived with the plant for years. It is more common in the spring when the young shoots are more palatable. In upland cattle grazing areas, herds are known to live quite happily alongside the bracken, then without warning will consume too much of it and be poisoned. If you have a lot of bracken , seek local advice.

Do your research and make sure you can identify these plants to avoid a problem in the first place.

Bloat

'Ordinary' bloat is serious, but if treated in time is seldom fatal. It is caused when the animal's main stomach (rumen) fills up with gas. Further study of this ailment is advised (see Further Reading). Frothy bloat is far more serious, as the name suggests, the gases form a froth with the contents of the rumen, (see below for further discussion of this ailment).

With bloat **prevention is the best form of cure**. Before turning animals out on to very lush pasture provide them with hay or straw to ensure that they have ingested enough fibre and do not have the internal capacity to gorge themselves, particularly if they have been used to poor grazing. Limit the amount of the new grazing they have access to by electric fencing, and/or, limit the time they have access to it. The presence of too much clover increases the risk of bloat considerably. Manure spreading in the winter helps to maintain the correct balance.

Treatment: Ask your vet for instruction on emergency actions that you can take if he is unable to attend to your animal quickly. Peanut oil, also known as groundnut oil, and washing soda crystals (in a water solution) can be used as a first aid remedy by drenching as described above (see Hypoglycaemia), but it may be more effective to carefully insert a length of piping down the animal's throat, (thin hosepipe in an emergency) to get the fluid straight to the stomach. Ordinary bloat can be dealt with in a different way by piercing the abdomen on **the left-hand side** at a specific place with a special tool. You must seek professional help and instruction before attempting this for the first time otherwise you may cause irreparable damage.

Frothy bloat

Symptoms and causes: Frothy bloat can be worse than 'ordinary' bloat in as much as it can be more difficult to save an affected animal. Frothy bloat is more likely to occur if there has been a long dry spell, followed by rain. The pastures will have been semi-dormant due to the dry period but the rain caused the grasses and clovers to grow very quickly. This rapid growth causes the plants to hold high levels of nitrogen and water, but few other minerals and little fibre. The animal will consume it with relish after a winter of dry fodder, but does not produce enough saliva to digest the plant matter properly. A foam forms along with the normal digestive gas, the animal swells, particularly high up on the left-hand side, and it cannot belch or pass wind as the stomach is completely blocked off by the action of the foam, which is made up of minute bubbles.

The area eventually becomes distended and tense. The animal may kick its sides or belly in obvious discomfort, its ears may be back and its breathing increasingly laboured, in the late stage a nose bleed may occur. Eventually the animal will suffocate as the swelling leaves no room for the lungs to work, causing a horrible painful death. This whole process can be over in an hour and so. If it happens at night you will not even have a chance to intervene.

Remedies and prevention

If you are on old permanent pasture you are at less risk of frothy bloat. However, if you have young grass leys for nitrogen fixing with broad-leafed clover, especially red clover, then there is real danger. If you think your pastures may risk causing frothy bloat, restrict the grazing with electric fencing and provide palatable hay or straw in the field, the cattle will consume it alongside the lush grass to their own benefit. If the animals are housed at night, make sure they come out in the morning full of hay or straw.

If turning cattle out on to fresh spring grass, you can let them have access to anti-bloat licks a few days before These licks can also be provided in the field as well (also dry fodder as above). An old natural remedy is to pour peanut or linseed oil into the water trough as a preventative measure, but you can never be certain how much each animal will drink if the grass is full of water.

Note: DO NOT PIERCE THE ABDOMEN IN AN ATTEMPT TO CURE FROTHY BLOAT IT WILL NOT HELP IN ANY WAY. If you suspect an animal has frothy bloat it needs to be dosed (drenched) with peanut oil made up of 1 pint (0.57 litre) of peanut (groundnut) oil in some warm water or, a quarter of a pound (113g) washing soda crystals dissolved in hot water and diluted to approx one pint (0.57 litre) with cold water. You will need help to do this, do not hold the cow's nose as this will make it more difficult for her, pass your arm over the nose and calmly and carefully put your hand in the mouth. It is amazing how quickly their stomachs will deflate after drenching. As always call the emergency vet as the animal may need a tube to be inserted into the first stomach to let any gas escape. This should only be attempted by the vet or by someone who is experienced unless it is literally a matter of life or death and help is not available immediately.

Mastitis

Symptoms: inflammation of the udder. You will be able to see that an infected teat looks different, probably swollen, it will feel hotter too. Check each quarter of the udder carefully, it should feel soft and pliable unless engorged with milk. In bad cases the udder feels hard and hot and looks inflamed. To aid diagnosis you could try and gently squeeze some milk out, it may contain clots and look puss like, a sure sign you have a problem. Mastitis is more common in milking cows, but it can also be a problem at weaning and particularly during the summer with suckler cows if teats become damaged. Make your cows' udders part of your regular routine inspection when they are in milk. The cows will become quite used to this as long as you avoid any sudden movements.

Treatment: if you suspect a problem seek advice from your vet as soon as possible. The condition is very painful for the cow, the calf will be prevented from getting its full ration and permanent damage can be caused to the cow and her calf.

Chlamydia

Better known as a sexually transmitted disease affecting humans. What is less well known and publicised is that it is probably a serious problem for domestic farm animals as well, though not through sexual activity. In fact, it could be the underlying cause of many farm animal ailments but this is not well documented as yet, to learn more, go to www.ela-europe.org, and don't forget, you heard it first here!

Notifiable and infectious diseases

There are a number of diseases which are notifiable by law. This means that you must inform your local animal health department if any of them have been diagnosed on your holding. Your vet will advise you of the correct procedures. In addition, in the UK, as part of BSE control measures, the agriculture ministry (DEFRA) monitor all on-farm deaths of animals greater than 30 months of age. These animals will be analysed for the presence of BSE as part of a national monitoring programme. This service is free.

Below is a list of some of the main notifiable diseases in the UK.

> Foot and Mouth disease
>
> anthrax
>
> warble fly infestation
>
> bovine spongiform encephalopathy (BSE)
>
> tuberculosis
>
> brucellosis (Z)
>
> enzootic bovine leucosis (EBL)
>
> Johne's disease (Z)

Key (Z) denotes zoonoses see below

Zoonoses

These are animal diseases that can also be transferred to humans. Some of the notifiable diseases which are also transmissible to humans have been eradicated in the UK, for instance, the last case of brucellosis confirmed in the UK was in 1993. It is however incumbent upon you to be aware of these particularly dangerous disease threats as a responsible animal keeper. Rather than include a detailed doomsday list here, you will find more detailed information available from the Further Reading section. Visit the DEFRA website, accessible for free from your local library, at www.defra.gov.uk/animalh for up-to-date information. On a global level, wherever you are, go to the website for the World Organisation for Animal Health at www.oie.int .

E. coli

A generally misunderstood threat to you, your family and, most likely, any visitors to your farm, is E. coli. There are harmful strains of this otherwise benign bacteria which cause potentially severe stomach upsets and illness and so it does have to be taken seriously. Outbreaks of E. coli poisoning make the news quite regularly and in a rather alarmist fashion. This in some cases has had the unfortunate and unnecessary effect of schools ending visits to open farms. One theory amongst livestock keepers is that visitors who do not have regular contact with animals may be less resistant if exposed to E. coli bacteria.

Prevention is simple however – sensible hygiene precautions must be observed at all times and this includes you! If you have touched the animals, have been mucking out, or have been sprayed with slurry or urine, if you were in the wrong place at the wrong time, wash your hands and any other part of you that has been affected, as soon as is practical. Keep your hands away from your mouth.

As far as farm visits are concerned, in the UK, your local Health and Safety authority will have their own rules and guidelines for organised visits which you will need to implement, which include simple things like the availability of washing facilities, wet wipes etc..

Notifiable diseases

BSE (Bovine Spongiform Encephalopathy)

Symptoms: Muscle tremors, aggression, staggering, and loss of use of the back legs. This is not to be confused with hypomagnesaemia and other conditions with similar symptoms as you are more likely to come across these latter.

Action: contact your vet immediately on seeing the above symptoms. This is a notifiable disease. In the UK when acquiring stock you can find out the BSE history of the source farm from DEFRA.

Foot and Mouth disease

Symptoms: affected milking animals show a sudden drop in milk production, together with blisters in the mouth, slobbering, sucking, high temperature, discomfort and pain when walking, blisters between the claws. Report these symptoms to your vet immediately, this is an extremely contagious, legally notifiable disease.

Tuberculosis and brucellosis

Your herd will be tested for TB and brucellosis at the expense of DEFRA once every two years. The herd tests are co-ordinated and carried out by your vet. You do need to be aware that some parts of the country are more badly affected than others by TB and you should do all you can to minimise the potential risks to your herd. Send for information from DEFRA. Tuberculosis and brucellosis are notifiable diseases. If buying in cattle from a known TB 'hotspot'– have the animals tested before delivery. It makes no sense to import infected animals into a clean area

TB is probably the greatest health threat of all to cattle in the UK. Study the issues carefully and stay informed. In the UK currently the debate rages on about the causes and methods of transmission and whether or not badgers are involved in transmission.

Other ailments and diseases

The following is a list of some of the other ailments and diseases you should be aware of. Most of these are extremely uncommon in the UK, however further research via the Further Reading list is essential.

Husk; fog fever; choke; big leg; black leg; redwater (particularly in Wales, Scotland, South West England, Devon, Dorset and northern England, this is a tick-borne disease); tetanus; leptospirosis; blaine; photosensitisation; wooden tongue; viral pneumonia; IBR; catarrhal fever; bovine viral diarrhoea; salmonellosis; liver fluke infection; mud fever.

This is not a complete list and does not include tropical diseases for readers in such areas.

Administering medicines

You need to keep first aid equipment to hand to enable you to give simple medication without calling the vet. Instruction for the method required for the different medicines is on the bottle or pack, read the instructions carefully to be sure you are choosing the right option and dose.

Ask your vet to provide you with the medicines you should have to hand which cannot be purchased from your agricultural merchant.

Your vet will also be able to give you instruction on how to give basic injections and how to dose animals orally (drenching). There are three types of basic injection:

1. subcutaneous: the needle slides sideways under the skin but does not pierce the muscle, the medicine rests under the skin and is absorbed slowly, it is the least painful injection for the animal, unless you are administering a solution which stings.

2. intramuscular: the medicine is injected straight into a muscle, your vet will demonstrate the best method.

3. intravenous: the needle is placed into a vein. This is a potentially life-threatening procedure if attempted by the inexperienced, on no account attempt intravenous injections without having had proper training.

In addition you should keep the following:

antibiotic spray for wounds, sore hooves, naval dressing

penicillin – injectable solution

Oxytetracycline – a broad-spectrum antibiotic in the form of an injectable solution

You can obtain the following items from your agricultural merchant.

thermometer (digital ones are easy to read and are not too expensive)

Scour (diarrhoea) **formula** (particularly if calf rearing)

colostrum substitute

liquid iodine, for wounds and naval dressing

disinfectant for washing the operative's or vet's hands as well as for cleaning surfaces or instruments

magnesium and calcium solutions, for staggers (calcium/magnesium deficiency). You must always keep this in stock, and make sure it is within its use by date. If you find yourself in an emergency and you do not have any to hand you will probably be too late.

obstetric gel for assisted calvings

16 gauge x1 inch **disposable needles**

disposable syringes, mostly 20ml but also a few 50ml

50ml **dosing** (drenching) **syringes**

bandage and **wound dressings**

calving ropes

stomach tube kit

weak calf reviver

Chapter Twelve
Showing, Shows and Sales

Shows present the best opportunity to promote the breeds, the breeders and their herds.

They are usually excellent social occasions, bringing like-minded people together from far and wide, like an annual reunion. You will find that the majority of show people will be only too pleased to help and encourage you. The main showing season runs from May to as late as September.

While some breeders can be fiercely competitive the majority of exhibitors take part for the best reasons and enjoy the experience of taking part, whether they win or lose. They also know, that their animals are good in their own opinion, the winners are simply the best as far as that particular judge on that particular day decides.

As far as the conservation of breeds is concerned, showing can have a negative effect. Size is generally of great importance in showing – an imposing animal is halfway to winning a prize. That said, shows are primarily for animals which have been specially selected to demonstrate their beauty and potential. There is pressure, albeit to a lesser extreme than in the commercial sector, to select for desirable show characteristics, but to date the negative effects have been mainly noticeable in the sheep breeds. Judges specialising in rare and traditional breeds are far more likely however to appreciate the natural functional attributes of an animal rather than just the fact that it has been fed and prepared well.

Highland cow champion at the Great Yorkshire Show.

Local shows

Your small local show is the first one you should consider showing at. It will probably be a one-day event and you won't have to travel far.

Even at a relatively small event you will be publicising your herd and your special breed. You may attract buyers for your surplus breeding stock, and meat or dairy products if you produce them. Many shows have young handler classes, so there is an opportunity to involve children

It is good to show your support at your local event, entries are always welcome and the organisers who commit so much time throughout the year deserve plenty of support.

Young handler class at the Three Counties Showground, Malvern.

National and regional shows

large-scale agricultural shows, such as the Royal Show, at Stoneleigh, or the Great Yorkshire, may last for two or even up to four days. In the UK some shows, like the Royal Bath and West, have a rare breed cattle show on the last day only, which makes it possible for more people to attend as they need only find one day from their busy schedule.

Initially, it can be an excellent idea to see if you can help an exhibitor at one of the large shows before you commit yourself to attending in your own right, often exhibitors are short of handlers because they are showing animals in different age groups and classes which can overlap. Helping in this way will give you invaluable experience and confidence. You will learn handling skills, but also how to wash, groom and prepare animals for showing. You will also pick up a few 'trade secrets' about preparing animals many of which are breed specific, for instance, there are ways of getting white animals to look superwhite on the day and there are 'kitchen' remedies for cleaning soiled coats and so on.

Combined Shows and Sales

Sometimes shows are combined with sales, particularly within the rare breed movement in the UK and they are an excellent opportunity for buying and or selling, providing you take sensible steps to isolate any purchases you take home for safety's sake. A sale is usually held at a market, but it is on a smaller scale with fewer sellers and animals than on commercial market days.

Card grading

At rare/traditional breed shows and sales some breeds implement a system known as card grading. This is a form of breed inspection and is different to show judging. The animals are inspected prior to the sale by breed experts, who assess the animals individually in line with established breed standards and physical functionality (colour, good teeth, two testicles, good udder, conformation/size etc..). The results are attached to the pen. Different breeds have different systems but usually coloured cards are used. The aim is to help buyers, particularly novice buyers, choose breeding stock with some confidence. It is a controversial procedure as far as some breeders are concerned, and is mainly applied by sheep breeds, but the benefits have been demonstrated and it is a useful guide for beginners.

Show classes

Classes are usually divided into different age groups, for example, cow or heifer born before 31st December 1999, bull born before 31st December 1999 and so on. There are also 'fatstock' or finished cattle classes for meat animals which sometimes take place under cover in the winter. Class categorisations do vary from event to event. Try to acquire a copy of a show schedule from a previous year from the show society or one of your fellow breeders, it will provide you with a clear idea of the classes and events at your chosen show.

David Powell's traditional Hereford bull, supreme champion at the Rare and Native Breeds Cattle Show, Three Counties Showground, 2004.

Bear in mind that there may be no separate class for your breed due to a lack of numbers. In this case you will have to enter the 'Any other breed' class and compete with other breeds in the same position.

Once you have decided which animal(s) to show and in which class, your work starts immediately, in a way, you cannot start too soon. At this point it will pay you to discuss your plans with someone with experience of showing who can give you breed-specific hints and tips. For example, it may be that to ensure your animal has a perfectly smooth coat for the show, you will have to 'rug it up' for a certain length of time to ensure you get the perfect coat at the right moment. This means putting a coat on the animal, made of jute sacking or similar, to help protect and control the condition of the coat, seek advice on the need and timing. This practice is most common on smooth-coated cattle such as Aberdeen Angus, it also helps to keep flies off. If a good coat of hair is required on, a beef steer for example, you would not use a rug as it limits the growth of hair.

Once the warmer weather arrives it would also be advisable to wash the animal say once per week to get it used to being pampered. Most animals greatly enjoy being brushed, and this helps them to enjoy the whole process right up to and including the show. You should also bear in mind the condition of the animals' hooves and plan well ahead, you will not be able to take remedial action close to show time as any work on the hooves may make the animal lame at the worst possible time.

Health requirements and animal movement rules

You will need to be aware of health requirements and any animal movement rules that prevail close to show time. There are national rules governing this (DEFRA) but local restrictions may also apply. This information will be in the show schedule which you should apply for in early spring.

Inspect your animals carefully before you set out, as they will be inspected thoroughly by vets on arrival at the show. For example, it is not uncommon for animals to have warts and it is as well to be aware that these animals will be excluded from the show in the same way as animals with other contagious conditions such as ringworm. In addition you are likely to need to provide the show organisers with a health declaration from your own vet. All of these details will be set out in the show schedule.

Showing dairy animals

Showing dairy animals is much more specialised than exhibiting beef animals. For instance the shape and fullness of the udder will be a key factor in the judge's mind, milking out the udder before entering the ring, to ensure it is perfect at the right moment is an art. Novice showers should look for as much help and advice as possible until they have gained sufficient experience of their own.

Judging

The animal must be relaxed and easy to handle on a halter and you will also need to be able to get your animal to stand to its best advantage too. This has to be practiced at home, with the help of a showing stick, obtainable from specialist suppliers (as listed in the Contacts section) but to start with a bamboo stick or walking stick will suffice.

The showing stick is used to prod the animal's hoof to get it to change position. You should repeat this until you are happy with the result. It is obvious that you cannot leave this practice until the day you are in the ring, your animal must be used to the

A typical, tall, polled Modern Hereford bull containing imported bloodlines. (Photo, Farmers Weekly. Courtesy of the Rare Breeds Survival Trust.)

process. The show stick is also used to stroke the belly of the bull near his sheath, which needless to say he finds pleasurable and stops him from feeling bored or fractious.

There are a number of general characteristics that a judge (or buyer) will be looking for whichever class you have entered. Above all, the animal must be quiet when led, paraded and inspected in the ring and it should not show any signs of lameness. I, like many judges, try to be as understanding as possible when animals misbehave, and make every effort to encourage exhibitors, but you may as well aim for the highest standards. All too often I hear that people have fetched the animal straight out of the field that morning, hosed it down under protest, with predictable results in the ring.

Showing etiquette

There is a long tradition of showing etiquette which you should be aware of. You should not try and chat to the judge while he or she goes over your animal and certainly do not try to influence the judge by explaining that your animal came first at the last show!

Historically judges are under no obligation to explain why and how they made their decisions, but an increasing number do and you may be able to learn a great deal from listening to them. I believe this to be a very positive development.

It is also best to remember that the judge's decision is final, their task is very difficult at most shows as the standards of the animals are very high, accept the decision with good grace, try and learn something from it, and look forward to your next opportunity. You must also consider your own appearance, both you and your animal should be perfectly groomed! A clean white coat can cover your everyday workwear very successfully.

Tidy lines

A number of shows award a prize for the tidiest place in the accommodation lines. It is essential that muck is cleared away in good time in the mornings, and that a high standard of cleanliness and tidiness is maintained throughout the show. Some exhibitors also provide informative displays about their breed which is an excellent idea.

Preparing an animal for showing

In Chapter Five we discussed halter training for management purposes, for showing you will also need to put time into teaching your animal to walk by your side on a halter. Short regular lessons accompanied by a reward is probably the best method, and it may help if you have an assistant who can encourage from behind. This method is time consuming, but it is the kindest way to build up a good relationship with your star animals. Preparing for showing and the build up to the show will be more enjoyable if the animal can be encouraged to cooperate. It will also make showing easier as the animal will be more relaxed.

You will need to keep your show animals separate from the rest of the herd, at least for some of the time. This also presents the opportunity to get them used to being tied up for increasingly prolonged periods, which is what will happen at a show. Again, link this experience to feed and some fuss and grooming, it may take a little time but in the end they will enjoy all the attention.

Get your animal used to being washed. This is best started when the weather begins to get warmer in the spring, it is not best achieved on a winter's morning with ice cold water! It is worth noting that hosepipes can scare the animals so be as sensitive about getting them used to this procedure as you have with everything else. Pressure washers are increasingly being used, but great caution is needed to avoid injury to the animal.

In order to get animals used to being transported to shows in a trailer it can be helpful, as show time approaches, to park the trailer so as to make it part of the animals' pen. Place plenty of clean straw in the bottom to encourage them to go in and out of their own accord. It reduces the stress of travelling, which is especially useful if the animals have never travelled before.

Many people take a limited supply of water from home with them when they go to shows, together with a supply of the hay and feed the animals are used to. You can usually buy hay on site, but any change in their usual feed or water can affect animals adversely and as a result they will not look or feel their best.

Feeding the show animal

In some herds calves are treated differently from the rest of the animals as soon as they are born, to prepare them as potential show animals with feeding. However, you would be wrong to think that stuffing an animal with food will produce a show winner. You need to ensure that you use the best

Eileen Hayes showing Dexters. (Photo, Eileen Hayes)

conserved forage, and other highly palatable feed to give yourself the best chance. You need to feed to condition which is an art, but one you will quickly feel competent at – it is all in the eye.

Traditional breeds can be brought up to show condition more easily than some of the more 'modern' breeds, however, it takes time to alter an animal's condition. It is inadvisable to do it too rapidly as sudden changes can affect the animal's health.

It is possible to acquire 'condition scoring' charts against which you can compare your animal, but most breeders do not use one.

Qualities and traits

In showing, an animal should neither be too thin (poor), nor too fat. With beef animals, the judge will want to see the animal's potential for having meat in all the right places, but don't over do it, if it wobbles all over with fat it will not do well. In dairy animals, amongst other things such as overall shape, the condition and shape of the udder at the time of judging will be very important. Dual-purpose breeds will be considered from both a meat and a dairy point of view. But in all classifications it is the following attributes that a judge will be looking for.

Type

Within breed types there are individual types and their popularity can be influenced by fashions within a breed. You will soon find out which are the current favourites whilst going through the process of establishing your herd. The starting point for you to learn the details of the ideal characteristics of your animal is from the breed society who can supply you with clearly defined descriptions

> 'An ideal or standard of perfection which combines all the characteristics that contribute to an animal's efficiency for the purpose specified.' (Selecting Fitting Showing Beef Cattle, Nordby Lattig, Interstate, Illinois USA, 1948).

Breed type

These are the characteristics which distinguish one breed from another, refer to the society breed description for specifics and make yourself aware of potential breed 'faults ' such as mis-marking. Remember it is possible for an animal to possess most of its breed characteristics yet not be of a suitable type! It might be lacking in some particular way.

Breed character

This can include colour, refinement as opposed to coarseness, style, carriage (the way it carries itself) and what I can best describe as 'presence'. A beef animal will immediately look as though it has the perfect potential for producing meat, its main function for representing its breed. A dairy cow will look as though she is capable of producing lots of milk, and she will be feminine looking,

Balance

This refers to the symmetry of the animal, in other words the harmonious development of all parts of the overall beast.

Quality

Quality describes the shape of an animal as well as how closely it fits the purpose that the breed was developed for. The opposite of quality is described as 'coarseness'. Quality should not be confused with refinement as a refined animal may be too light-boned to be a good beef animal for instance.

Scale

This refers to how well developed the frame of the animal is and to its overall size. The frame must be properly developed before you start to worry about fleshing and condition, if the animal's frame is below the average for the breed, no amount of care in the feeding for show will solve the problem.

Substance

In beef cattle this refers to the amount of bone and the overall ruggedness of the beast, obviously dairy cattle are judged on different parameters.

Style

This term refers to the general beauty, demeanour and carriage of the animal, in other words how they walk and behave in the ring. This attribute can take your animal further than its physical equal which does not present itself as well.

Finish

This refers mainly to the degree of fatness of the animal. The animal should be adequately 'finished' with an even distribution of fat, but it should be firm to the touch, not flabby. Perfection in this category is the result of mastering the art of feeding.

Red Poll bull, Tatton Red Eric, breed Champion at the National Rare Breeds Show and Sale 1993.
(Photo, Simon Tupper. Courtesy of the Rare Breeds Survival Trust.)

Constitution

This terms mainly refers to chest capacity in the case of beef animals, and judges will be looking for length, fullness and width.

Final preparations for showing

Make a checklist of all you will need for the show. This should include fodder (remembering that too much concentrate may mean extra washing duties!). Straw is usually provided at shows for bedding but it can be a good idea to take extra. Show equipment suppliers listed in the Contacts section can also offer advice as well as other breeders.

Show box

You will need a show box to hold all your equipment. This could be a simple plastic crate with a lid, or an old pine chest, which could have the herd name painted on the lid.

Below is a list of recommended contents: grooming brushes, shampoo, showing stick to help adjust the animal's stance, sponges

Also, enough white coats so you can look smart yourself.

Allow for the worst weather conditions; a hat is also useful for prolonged exposure in the sun.

If the cattle stalls have overhead fittings, you will be able to display a special sign specific to your named herd. These signs may include simply the herd name and address with contact details of the exhibitor, or may be more elaborate and include an illustration. Signs make a good impression and can help attract potential buyers.

Beef Shorthorn Bull, Rare Breeds Survival Trust Show and Sale 1997. (Photo, Eileen Hayes).

Conclusion

I hope you will show your cattle; apart from the promotional benefits that will flow for your business, it's great fun and is a strong part of the social element of cattle keeping, which is very special in itself.

Chapter Thirteen
Marketing

This vital activity covers two areas, livestock and meat. Without income, keeping cattle is a relatively expensive but rewarding hobby.

Livestock

One of major advantages of choosing pedigree registered stock is that it gives you access to a premium market for selling animals with breeding potential because they are unusual compared to 'run of the mill' cattle. If you are concentrating on special breeds, you can become well-known within your breed quite quickly if your stock is good. This in turn can insulate you from the normal market prices for surplus unregistered stock which either goes to finishers (fatteners) or straight to slaughter, usually via markets or marts.

If the price happens to be low at the time you need to sell (when the animals are ready they are ready!), then it can be a more viable option to keep breeding stock until you interest the right buyers. If you are unable to find a buyer at the right price and you have the space you can keep a few more for yourself. In the UK, pedigree stock have proved to be a sound investment over the medium to long term. At times when there is a shortage of quality breeding stock then prices are excellent.

To achieve a good name as a breeder of fine animals your skills at promotion will be called into play. Showing (see Chapter Twelve) is the obvious starting point, as on these occasions you will be able to display your goods with pride and to the perfect audience. You will get a lot of feedback. If keeping rare or traditional breeds it should be easier to get other publicity too because the animals are unusual and always generate interest.

Not everything you breed will be suitable to breed from and so will have to go for meat. With bulls in particular you will need to be very critical. The market for live bulls is limited, especially as so many people use artificial insemination. Breed societies usually keep a register of stock for sale and this should be your starting point.

There are number of interesting ways of generating publicity locally which are also likely to attract potential buyers. Open days and herd visits are popular events. It may well be that even during the hardest times you will be able to generate sales of your specialist stock, particularly if they are kept in good condition. However, if financial constraints mean a less generous feed regime, most of the breeds being advocated in this book will regain condition quite rapidly when required, although the larger traditional breeds may not be quite as easy to get back into condition.

Meat

It is uniqueness which is the greatest selling point for meat from pedigree or purebred stock, as it is for livestock. However, good meat animals must be killed and processed correctly if their full potential is to be exploited. A stressed animal which has travelled a great distance, even when killed humanely, is unlikely to result in the highest quality meat, particularly if it has had to wait around with lots of other strange animals in a strange place. You will need to maintain complete control of that process to ensure your best chance of success. If at all possible (increasingly difficult in the UK) use a small local abattoir to achieve this.

You will need to bear in mind though that individual animals can vary in terms of the quality of the meat they produce – you will be on a steep learning curve as you start out. Advice from more experienced people throughout all the stages of livestock and meat production is invaluable.

The cash flow for beef production is not good. It takes at least 18 months to finish an animal by the methods I am recommending, and it is difficult on a small scale to arrange calving so that you have a beef animal ready say once per month. That said, you can use the 'seasonality' of your special product to enhance its desirability – the French have proven how successful this can be with their Beaujolais Nouveau wine initiative each year.

Marketing and selling

Outside the UK, the rules for meat marketing are likely to be more relaxed. Within the UK, the Internet has to be your first port of call even before your first calf is even born. You can research all kinds of marketing initiatives at both home and abroad to develop and fine tune your plans and ideas long before your first sales are to take place.

Selling your meat directly to your customers can be time-consuming and may require you to have some knowledge of marketing, but the advantages are that the profits will be yours and you will benefit in all areas by making a name for yourself. Developing your own unique outlet can also be time-consuming, expensive and will be severely restricted by regulations and legal requirements.

An easier option may be to collaborate with a local specialist butcher whose work will be of a very high standard, but this does come at a price. In that context it is reasonable for you to charge a premium for the meat, as long as it meets the high standards that customers would expect for the extra money they are spending. If the butcher in return is guaranteed an exclusive supply of gourmet quality meat the arrangement can work well for both parties.

There are also a number of other options you should consider.

- Farm shops have appeared all over the UK, one of those near to you may be very interested to hear what you are doing.

- Farmers markets both in the UK and abroad are enjoying phenomenal success, but gearing up to participate is a big commitment in terms of both time and money. In the first instance, see if you can find an established market trader who might like to work with you.

- Restaurants are another possibility, but they tend only to want specific cuts (the best) which you could sell many times over anyway. To give you some idea, if you imagine the silhouette of a cow and draw a line from the top of the shoulder to the bottom of the thigh, all of the meat above the line will sell faster, it contains all the favourite joints and steaks. Try and persuade your restaurant customers to agree to take a proportion of the cheaper cuts as part of the deal. Specialist restaurant suppliers are also worth seeking out. One supplier I have worked with closely in the UK has always gone the extra mile to look after those who provide his produce, as a result, his business and his producers work on a stable basis. Look for this kind of specialist.

General notes on marketing

1. Marketing need not cost a fortune. Your special product combined with your enthusiasm and commitment are powerful marketing tools. You have a product you can feel proud and confident of. Even if you have never thought of yourself as a salesman before, it can be done successfully by anybody with a little passion for what they are doing

2. By all means have an inexpensive presence on the Internet, but do not expect a wave of interest, if you want a high profile on the net, you need to spend a great deal of time and some money to drive visitors to your site. Lots of specialist producers do sell direct this way, seek them out, but try supplying them instead of blazing your own trail early on. Setting yourself for this type of enterprise costs a great deal of money, and the margins are narrow.

3. There are organisations that can help you with marketing, farm assurance schemes, organic certification etc.

4. Recipes, both traditional and new can attract attention to you, food magazines and programmes are always looking for something new or interesting.

5 Packaging, from the simplest leaflet to accompany a cut of meat to more elaborate and interesting packaging, if you are brave enough to go down the food preparation route, it will make a difference. If you supply a butcher, provide him with A5 leaflets which are cheap to produce, to go with each item of yours that he sells. Describe your natural production system, some history about your breed, perhaps include a recipe, let them know how much care has gone into bringing that customer a gourmet item. You can even include information about any other conservation activities taking place on your farm. You would be surprised how much of this information will reverberate around the customer's dinner table, no doubt resulting in further orders.

Processing and storage

In advance of your first load of beef arriving, in the UK contact the Meat and Livestock Commission (MLC) and acquire their very helpful information on the different cuts of meat you can get from a carcase. They are there to help you, every time you have an animal killed you pay a levy to help fund them!

Assuming that you are having your animal(s) killed at a local abattoir in the least stressful way as described above, the following procedure is the only one allowed in the UK without having premises licensed for storing and cutting meat. Once the animal is killed and skinned, it will be cut in half and subsequently into quarters. The forequarters contain the cheaper cuts for stewing etc. and for the production of minced beef, the hind quarters are where the prime cuts are.

The meat needs to be 'hung' (matured) for on average 21 days in a proper cool room at the butcher's, where they regularly monitor the storage temperature and conditions, including humidity. Storage capacity is often a problem in smaller establishments so you may have to compromise on the length of time your meat is stored, younger animals may not need as long.

Modern meat production methods rarely allow for this vital phase of meat preparation. Bear in mind that your market is quite different from the mass market and it is the maturation period which will produce the top quality meat for your more exacting market.

When it comes to cutting the carcase up, if it is for your own freezer it is relatively straightforward. You need to decide what size roasting joints you want, how thick you want the steaks cut, how many per pack etc.. Similarly for the forequarter, have a proportion as braising steak, the tough parts cubed for stewing steak, and a proportion into packs of minced beef .(you can specify how lean you want this to be). I would also recommend, if your butcher can do it, to make some into sausage, in half kilo packs. It is a very good idea to have most of the meat vacuum packed on his premises. Some people do not like vacuum-packed anything, but it is the cleanest and safest option for delivery to retail consumers.

Delivery

When delivering meat you have to practice what is known as 'due diligence', this means that the cuts of meat must not reach the consumers at above 4°C. Apart from selling from his shop, the butcher may deliver to his customers in a fridge van. An option you could look into is a portable fridge box which plugs into your vehicle lighter socket, but by necessity they are quite small and not cheap.

USP – Unique selling points

There are a number of advantages and unique selling points which you can emphasise to help market your meat. These are of particular interest to a sizeable sector of the meat-buying public. People have never been more aware of or interested in how their food was produced, particularly if it has been produced in a way they approve of and is of excellent quality.

The following points are just some of the advantages of your product.

1. The meat has come from an animal which has enjoyed the highest welfare standards, it has matured naturally at its own pace, and has been fed largely on grass, which is its natural food.

2. Traditional and rare breeds generate interest in themselves, and all have a fascinating history.

3. The meat may have been produced organically.

4. It is a local product being sold locally – traceability.

5. Grazing stock on grass leys which contain a rich variety of herbs adds to the flavour of the meat (pre-seasoned!)

6. High welfare standards and small-scale production will almost certainly mean little routine use of medication, such as antibiotics, and the use of mixed rations as opposed to bought-in compound feeds. These points can be highlighted even if you do not have organic status.

7. Finally, bring out the issue of the conservation and preservation of your country's or region's traditional cattle breeds.

Adding value

The items we have discussed above have already added significantly to the potential value of your product. If you are doing your own individual marketing you can take a further step with your butcher and produce specialities. Even if you limit yourself to selling freezer packs you will quickly run out of the choice cuts like rib and steak and other roasting joints. There is a limit to how much minced forequarter you can expect people to buy, so you need to make your packs as tempting as possible. Ask your butcher to produce beef sausage or beef olives and salt beef is another option for cheaper cuts like brisket. However, tempting the idea of producing pies, salt beef, sausages etc., from a commercial view it should be one of the last challenges you set yourself considering all the costly and detailed rules which apply.

Conclusion

One of your main priorities should be to retain control of what is essentially your brand, once it has left the farm gate. Your primary role, let us not forget, is that of a cattle breeder/keeper and so it is best to dovetail your production output into somebody else's established marketing outlet until you feel you have reached a point where you need a new challenge, or your herd has expanded to a point where you can justify the investment of time and money and some additional risk. One thing is certain from my own experience where we had to ration supplies, if you produce the best, you will be hard pushed to ever produce enough to meet demand – an enviable and enjoyable position to be in.

Chapter Fourteen
The Rare Breeds Survival Trust

It is the Rare Breeds Survival Trust (RBST) who have been the driving force behind breed conservation in the UK for over three decades, and their story is fascinating and inspiring.

The problems faced by the pioneers of what was to become the Rare Breeds movement were daunting. The dangers threatening wild animals and their habitats on a global scale were well known; domestic animals had some of their welfare interests taken care of by charities such as the RSPCA, but, largely unnoticed, another group of animals was beginning to disappear and there was no organisation in existence to stem their decline or even in some cases extinction. That these islands were responsible for the origin of over 50 distinct sheep breeds and numerous special local types was not common knowledge and that amongst these breeds were the direct descendants of hardy, self-sufficient and resourceful Neolithic sheep was not recognised by more than a handful of 'enthusiasts'.

One of the reasons for this neglect may have been that for many people, whales, dolphins, pandas or exotic birds are far more worthy of support than threatened breeds of domestic farm animals. However, unique breeds were disappearing which had been developed by man, and were specially adapted over many thousands of years for particular requirements, both environmental and production – meat and milk, skins for leather, fat for oil lamps, dung for fertiliser, and as draught animals for agriculture.

Human beings have proved very successful at selective breeding and over many centuries have produced increasingly productive animals. One of the undesirable consequences of this process was the gradual but seemingly unstoppable redundancy of traditional British breeds.

White Park ox on display.

Why save domestic breeds?

The case for saving the domestic breeds has gradually become as strong as that for wild animals. The main argument in both cases is the need to maintain biodiversity. If species disappear, ecosystems break down, mostly due to pressure from man, but also because of natural changes; food chains are broken, species and habitats disappear. Biodiversity, we are told means a healthier planet and healthier people.

Another key issue is genetic potential. There is much still to be discovered about the natural world, and each new discovery about an animal or plant often reveals benefits for us; this is equally relevant to domestic farm animals. Our increasing knowledge of genetics and a more open approach to evaluating the plant and animal life around us, have established the importance of maintaining our global genetic 'databank'.

The Rare Breeds Survival Trust

The Rare Breeds Survival Trust is the only national charity dedicated to conserving endangered breeds of British farm livestock and was set up over 30 years ago.

The formation of the RBST can be traced back, albeit indirectly to the founding of the Zoological Society of London in 1825. Its base was Regents Park in London. By 1925 the zoo was overflowing, with more than 5000 animals on display. The society established a second site for surplus or sick animals at Hall Farm, Whipsnade on the Bedfordshire Downs. Domesticated livestock was represented at both sites in a small way, but the importance of domestic breeds had still not been fully appreciated.

In 1955, Solly Zuckerman (later Sir) became Honorary Secretary of the society. On assuming his position he immediately began to address the 'doomsday scenario' he could imagine ultimately affecting domestic breeds. During the Second World War he had witnessed at first hand the indiscriminate slaughter of livestock across Europe both for food, and through the violent effects of war itself. Wild animal populations were not immune to the effects of war either. His vision was that Whipsnade could provide a safe haven for both wild and domestic animals, but with the latter on a much more comprehensive scale than the few sheep and cattle they already had. At the time, space was not a problem at Whipsnade, so the stage was set.

An unsubstantiated but credible story suggests that Zuckerman's views had been strengthened by the story of an extraordinary bird – the Scots Dumpy. Poultry at this time was being developed to fit into increasingly intensive production methods, worlds away from their traditional role. Battery cages were to be the future, and where would that leave the Scots Dumpy, with its short legs and renowned ability to forage for its own food? Zuckerman realised that the Dumpy was near extinction and while it could not have a place in the 'modern' poultry industry it may well have an unforeseen role at some point in the future, he helped to maintain this thrifty, dual-purpose breed. Zuckerman understood early on that future human generations should be allowed the widest choice of these unique breeds.

A significant development in the process of breed conservation was Zuckerman's involvement in helping to find a home for some imported Charolais cattle (which had been offered as a gift, having completed their role as research animals). Whipsnade seemed to be the ideal place to use as a holding unit. Little did anybody realise that at the same time some of our breeds were actually in the process of disappearing forever without any action being taken.

The following breeds had become extinct prior to the formation of the Rare Breeds Survival Trust.

Pigs: Lincolnshire Curly Coated, Cumberland, Yorkshire Blue and White, Small White, Oxford Sandy and Black (now revived), Dorset Gold Tip, Ulster White

Cattle: Glamorgan, Sheeted Somerset, Alderney, Suffolk Dun, Pembroke (Castlemartin), Caithness, Red Norfolk and Blue Albion (now revived)

Sheep: Rhiw, Cannock Chase, Morfe Common, Limestone and St Ronas Hill

Horses and ponies: Manx, Cushendale, Tiree, Long Mynd, Galloway and Goonhilly

The principles of conservation

In 1966 the Food and Agriculture Organisation of the United Nations (FAO) began to evaluate domestic livestock in terms of its value as a genetic resource. Their rather short-sighted conclusion was that domestic animals should be 'evaluated for their performance with respect to present or short-term future requirements'. This was a long way from the philosophy developing in Britain at the time and which was to prove an important spur for the conservation movement.

The original remit established by the founders of the RBST was as follows:

'For the benefit of the public to ensure the preservation of breeds and breeding groups of domestic livestock of importance in the promotion of agriculture, being numerically small and having characteristics worthy of preservation in the interests of zoological research and education, to ensure the preservation of genes of special or potential value in hybridisation and other work.'

Funding had to be found, publicity was needed to raise public awareness, and work began on developing the idea of a semen bank and other initiatives – the Rare Breeds Survival Trust was born

1990s to TODAY

The RBST brought the plight of endangered farm livestock to the farming community and public through a show demonstration programme which was a major commitment for the Trust. Two show demonstration units between them appeared at 60–70 shows a year, the show societies providing many facilities free of charge. Members local to the shows, together with approved farm parks and breed societies, loaned the stock, also free of charge. It was this hard work and the publicity it generated which began to halt the decline in our domestic farm breeds.

Rare Breeds Survival Trust promotional display at an agricultural show.

Unfortunately, the Foot and Mouth epidemic dealt a severe blow to this major part of the Trust's activity. Shows were cancelled, putting financial pressure on all those involved in them. The show circuit still has not yet fully recovered, newly imposed strict animal movement/bio-security measures also cause real problems; the free sites and facilities that the RBST had enjoyed for so many years have been drastically reduced.

However, the situation is improving, albeit slowly. A National Rare and Traditional breeds show and sale is now held in Melton Mowbray, Leicestershire, supported by the RBST. There are also several regional rare breed events aided by sympathetic local auctioneers. Three of the most notable are York Sale at Murton, Carlisle and Chelford in Cheshire. Events also take place at Salisbury, Lincoln, Exeter, North Wales and Skipton in Yorkshire. Dates for all these events dates are published in the RBST members magazine, *The Ark* and on their website (www.rbst.org.uk).

Against this background of recent setbacks a number of groups and individuals are making a real impact on breed conservation.

For many years the Prince of Wales has been the Trust's patron. On his organic farm in Gloucestershire he has a selection of pedigree rare breeds, including Irish Moiled cattle. The Princess Royal, who is also a very active supporter, has White Park cattle and Gloucester Old Spot pigs and has taken a keen interest in the Trust's work with rare horse and pony breeds. The Queen has Cleveland Bay horses and the Duke of Edinburgh has used Fell ponies in driving trials. They have all made a point of including visits to RBST demonstrations at shows part of their schedule.

The Trust's regional support groups, made up entirely of volunteers, have kept rare breed issues alive in their local area. Raising funds by selling merchandise and the Trust's profile through events such as school visits to rare breed farms. The volunteers come from diverse backgrounds, some do keep stock but others live in cities. These active local groups are the backbone of the RBST and provide an important arena for discussion and the exchange of experience and information.

The RBST is currently supporting the development of a national 'gene bank' for domestic farm animals which will function in much the same way as the national seed banks developed at a number of sites across the UK. The slaughter of many millions of animals as a result of the Foot and Mouth epidemic revealed just how important this initiative is.

The RBST website also contains a great deal of information on the history and background of traditional and rare breeds, on current initiatives and how you can become involved.

The future

As a result of trailblazing work by the RBST and its volunteers, public awareness has been raised and through a number of initiatives **no more breeds have been lost.** Similar organisations have been formed in other countries, most notably the American Livestock Breeds Conservancy (ALBC) in the USA and Rare Breeds International who as the name suggests, function on a global scale. The RBST continues to be the world's leading authority on rare breed conservation.

Changes in EU legislation for subsidy reform also present great opportunities for rare breeds as their importance is recognised on an international level. In addition, in the UK, collaboration with organisations such as English Nature and other opportunities for conservation grazing will almost certainly see a real renaissance for RBST breeds and their breeders.

In spite of this more positive outlook many of our traditional breeds are still rare and need a great deal of support. I hope that this book will encourage you to become part of their survival.

Contacts

UK Breed Societies

Aberdeen-Angus Cattle Society

Mr Robert Anderson
Pedigree House, 6 King's Place
Perth, Scotland, PH2 8AD
Tel: 01738 622477 Fax: 01738 636436
E–mail: inffo@aberdeen-angus.co.uk
Website: ww.aberdeen-angus.co.uk"

Ayrshire Cattle Society

1 Racecourse Road, Ayr, Ayrshire KA7 2DE
Tel: 01292 267123   Fax: 01292 611973
Email: society@ayrshirescs.org
Web: www.ayrshirescs.org
Contact: Mr David Sayce, General Manager

Beef Shorthorn Cattle Society

(inc Northern Dairy Shorthorn)
Mr Frank Milnes
4th Street, National Agricultural Centre
Stoneleigh Park, Warwickshire, CV8 2LG
Tel: 024 7669 6549 Fax 02476 696729
E-mail: milnes@shorthorn.co.uk"
Website: www.shorthorn.co.uk

Belted Galloway Cattle Society

Miss Myrna G Corrie
Parklea, Tongland Kirkcudbright, DG6 4ND
Tel/fax: 01557 820218
E-mail: myrna@lineone.net
Website:www.belties.com"

British Friesian Breeders Club

Contact: David Armett
Tel: 01530 223446, Mobile 07947 600418
E-mail: temraa@fwi.co.uk

British Kerry Cattle Society

Mrs Joan Lennard
Windle Hill Farm, Sutton on the Hill
Ashbourne, Derbyshire, DE6 5JH
Tel/Fax 01283 732377
E-mail errycattle@btinternet.com
Website: www.kerrycattle.org.uk

British White Cattle Society

Mrs Angela Hamilton
Southfield Road, Woodbastwick,
Norwich, NR13 6AL
Tel: 01603 722288 Fax: 01603 721168
E-mail: breedsecretary@britishwhitecattle.co.uk
Website: www.britishwhitecattle.co.uk"

Devon Cattle Breeders Society

Cattle breed society
Wisteria Cott, Iddesleigh
Winkleigh, Devon EX19 8BG
Tel: (01837) 810942 Fax: (01837) 810942

S D H B S; South Devon Cattle Society.

Office Suite 1
Westpoint, Clyst St Mary, Exeter EX5 1DJ
Tel: 01392 447494

Dexter Cattle Society

Ms Elaine Lester
Dishley Grange Farm
Derby Road, Loughborough, Leicestershire, LE11 5SF
Tel: 01509 646677 Fax: 01509 646688
E-mail dextercattlesociety@btopenworld.com
Website:www.dextercattlesociety.co.uk

English Guernsey Cattle Society

Produce herd book
The Gold Top Centre
Pednor Rd, Chesham,Bucks HP5 2LA
Tel: (01494) 774114 Fax: (01494) 778048

Galloway Cattle Society

15 Newmarket Street, Castle Douglas,
Kirkcudbrightshire, DG7 1HY
Tel: 01556 502753 Fax: 01556 502753
Email: info@gallowaycattlesociety.co.uk
Web: www.gallowaycattlesociety.co.uk

Gloucester Cattle Society

Mrs Emma Barber
41 Crows Grove, Bradley Stoke,
South Gloucestershire BS32 0DA
Tel: 01454 615715
E-mail: secretary@gloucestercattle.org.uk
Website: www.gloucestercattle.org.uk

English Guernsey Cattle Society

Scotsbridge House, Scots Hill
Rickmansworth, Hertfordshire WD3 3BB
Tel: 01923 695204 Fax: 01923 695215
Email: egcs@guernseycattle.org.uk
Web: www.guernseycattle.org.uk

English Guernsey Cattle Society

Produce herd book
The Gold Top Centre
Pednor Rd, Chesham, Bucks HP5 2LA
Tel: (01494) 774114
Fax: (01494) 778048

Hereford Cattle Society

Hereford House, 3 Offa Street, Hereford HR1 2LL
Tel: 01432 272057
Fax: 01432 350608

Traditional Hereford Breeders Club

Mr David Fenton
Honour Farm, St Michaels,Tenterden, Kent, TN30 6TJ
Tel: 01580 762395
Website: www.thcbc.co.uk"

Highland Cattle Society

59 Drumlanrig Street, Thornhill, Dumfriesshire DG3 5LY
Tel: 01848 331866
Fax: 01848 331183
Email: info@highlandcattlesociety.com
Web: www.highlandcattlesociety.com
Contact: Mr Ken Brown, President

Irish Moiled Cattle Society

Mrs Janet Kennedy
185 Craigs Road
Cullybackey, Ballymena, Co.Antrim
Northern Ireland, BT42 1PG
Tel: 02825880300
E-mail: janet.kennedy@ukf.net
Website: www.irishmoiledcattlesociety.com

Jersey Cattle Society of the Uk

Scotsbridge House, Scots Hill, Rickmansworth,
Hertfordshire WD3 3BB
Tel: 01923 695203
Fax: 01923 695203
Email: csoffice@jerseycattle.org
Web: www.jerseycattle.org
Contact: Mr Steve Baker, Executive Secretary

Lincoln Red Cattle Society

Mrs Lorna Newboult
Lincolnshire Agricultural Society,Lincolnshire Showground
Grange-de-Lings, Lincoln, LN2 2NA
Tel: 01522 511395 Fax: 01522 520345
E-mail: lincolnred@lineone.net
Website: www.lincolnredcattlesociety.co.uk

Longhorn Cattle Society

Mrs Sarah Slade
Southcott Farm,Chawleigh
Chulmleigh, Devon EX18 7HP
Tel: 01769 581212
E-mail: longhorncs@tiscali.co.uk
Website: www.longhorncattlesociety.com

Luing Cattle Society

Incheoch Farm, Alyth, Blairgowrie PH11 8HJ
Tel: 01575 560236
E-mail: secretary@luing-cattle.ndo.co.uk
Website: www.luingcattlesociety.co.uk
Contact: Judy McGowan

Red Poll Cattle Society

Mrs Terina Booker
52 Border Cot Lane, Wickham Market
Woodbridge, Suffolk, IP13 0EZ
Tel: 01728 747230
Fax: 01728 748226
E-mail: secretary@redpoll.co.uk
Website: www.redpoll.org

Shetland Cattle Herd Book Society

Miss Alison Bulley
Agricultural Marketing Centre
Staneyhill, Lerwick, Shetland Isles, ZE1 0NA
Tel: 01595 696300/692030 Fax: 01595 696305

Shetland Cattle Breeders Association

Mrs Mary Holloway
Barnack, Langley, Liss, Hants, GU33 7JR
Tel 0845 4588411 (lo-call) Fax 01730 895225
E-mail mary@barnack.fsworld.co.uk

Shorthorn Society

4th Street, National Agricultural Centre
Stoneleigh Park, Warwickshire CV8 2LG
Tel: 02476 696549 Fax: 02476 696729

Sussex Cattle Society

Station Road, Robertsbridge
East Sussex TN32 5DG
Tel: 01580 880105 Fax: 01580 881334

Welsh Black Cattle Society

Royal Welsh Showground
Llanelwedd, Builth Wells, Powys LD2 3NJ
Tel: 01982 551111 Fax: 01982 551333

White Park Cattle Society

Mrs Shirley Hartshorn
60 Rookery Avenue, Sleaford, Lincolnshire NG34 7TY
Tel: 01529 303958
Website: wwwhitepark.org.uk

Whitebred Shorthorn Assoc. Ltd

Ms Rosemary Mitchinson
High Green Hill, Kirkcambeck,
Brampton, Cumbria, CA8 2BL
Tel: 01697 748228
E-mail: rosiemitch@ukonline.co.uk
Website:www.whitebredshorthorn.com

Government and other useful organisations

DEFRA

Department for Environment, Food & Rural Affairs
Nobel House, 17 Smith Square, London SW1P 3JR
Tel: 020 7238 6000 Fax: 020 7238 6609
Helpline: 08459 33 55 77
E-mail: helpline@defra.gsi.gov.uk
Website: www.defra.gov.uk

Meat and Livestock Commission (MLC)

Website: : www.mlc.org.uk
MLC Website: : www.britishmeat-export.com

Food Standards

Websites: : www.food.gov.uk / www.foodstandards.gov.uk

Red Meat Industry Forum

Website: : www.redmeatindustryforum.org.uk

The Soil Association

Bristol House,
40-56 Victoria Street, Bristol, BS1 6BY
Tel: 0117 314 5000 Fax: 0117 314 5001
Email: info@soilassociation.org

The Soil Association Scotland

18 Liberton Brae, Tower Mains, Edinburgh, EH16 6AE
Email: contact@sascotland.org
Tel: 0131 666 2474,
Fax: 0131 666 1684

The Biodynamic Agricultural Association (BDAA)

Painswick Inn Project,
Gloucester Street, Stroud, Glos GL5 1QG
Tel/Fax: 01453 759501

The Rare Breeds Survival Trust (RBST)

National Agricultural Centre
Stoneleigh Park, Warwickshire CV8 2LG
Tel: 024 7669 6551   Fax: 024 7669 6706
Email: enquiries@rbst.org.uk
Website: www.rbst.org.uk

Rare Breeds International (RBI)

Website: www.rbi.it

American Livestock Breeds Conservancy (ALBC)

PO Box 477, Piltsboro, North Carolina 27312, USA
Tel: (919) 542 5704 Fax: (919) 545 0022
E-mail: info@albc-usa.org
Website: www.albc-usa.org

Rare Breeds Canada (RBC)

National Office
341–1 Clarkson Road, RR#1 Castleton, ON,
Canada K0K 1M0
E-mail: rbc@rarebreedscamada.ca
Website: rarebreedscanada.ca

British Association of Homeopathic Veterinary Surgeons

E-mail: enquiries@bahvs.com
Website: www.bahvs.com

Peasridge

Showing equipment etc.. for cattle
(Worldwide Service)
Stonelink, Stubb Lane, Brede, Rye, East Sussex TN31 6BL
Tel: 01424 882900
E-mail: info@peasridge.co.uk
Website: www.peasridge.co.uk

Organic Pasture Management

Elm Farm Research Centre
Hamstead Marshall, Newbury RG20 OHR
Tel: 01488 658298
E-mail: elmfarm@efrc.com
Website: www.efrc.com

Humane Slaughter Association

The Old School
Brewhouse Hill, Wheathampstead, Herts. AL4 8AN
Tel: 01582 831919
E-mail: info@hsa.org.uk
Website: www.has.org.uk

UK Agricultural Show Listings

Tel: 01730 266624
E-mail: info@showmans-directory.co.uk
Website: www.showmans-directory.co.uk

Traditional Breeds Marketing Co Ltd

Freepost (GL442),
Cirencester, Gloucestershire, GL7 5BR
Contact: Richard Lutwyche, General Manager
Tel/Fax: 01285 869666
E-mail:greatmeat@aol.com
Website: www.rbst.org.uk/htmlrare_breeds_meat.html

GAP (Grazing Animals Project)
Advice on Conservation Grazing in the UK
GAP Office
c/o The Wildlife Trusts, The Kiln, Waterside
Mather Rd, Newark
Notts NG24 1WT

Further Reading

Many of the following titles are available from Farming Books and Videos by mail order.
Visit www.farmingbooksandvideos.com

Practical and technical titles

Straiton, E., **Calving the Cow and Care of the Calf**, Farming Press, Ipswich, date Crowood, Marlborough, 2002 ISBN 0-85236-250-1

Straiton, E., **Cattle Ailments Recognition and Treatment**, Farming Press, Ipswich, date ISBN 0-85236-257-9 Crowood, Marlborough, 2000

Blowey, R., **Cattle Lameness and Hoofcare**, Farming Press, Ipswich, 1993

Day, C., **The Homeopathic Treatment of Beef and Dairy Cattle**, Beaconsfield Publishers, Beaconsfield, 1995

Coleby, P., **Natural Cattle Care**, Acres, USA (*I have not tried all the ideas in this book but it is worthy of your attention.*)

Thickett, B., Mitchell, D. & Hallows, B., **Calf Rearing**, Farming 1988

Sheldrick, R.D., Newman, G. & Roberts, D.J., **Legumes for Milk and Meat**, Chalcombe Publications, 1995

McDonald, P., et al., **Animal Nutrition**, 6th edn, Prentcie Hall, Harlow, 2002

Willis, M.B., **Dalton's Introduction to Practical Animal Breeding**, 4th edn, Blackwell Science, Oxford

Porter, V., **Caring For Cows**, Whittet, 1991

Phillips, C.J.C., **Cattle Behaviour**, Farming Press, Ipswich, 1993

Porter, V., **Fieldcraft and Farmyard 'Groundwork for Beginners'**, Swan Hill Press, 1992 (*information on fencing etc..*)

Van Loon, D., **The Family Cow**, Garden Way Publishing

Smith Thomas, H., **Storey's Guide To Raising Beef Cattle**, Storey Books

Boden, E., (ed.), **Black's Veterinary Dictionary**, 20th edn, A&C Black, London, 2001

Young, R., **The Secret Life of Cows**

Radford, **A Guide to Stock Fencing**

Conservation and History

Alderson, L., **The Chance To Survive**, 2nd edn, Pilkington Press, 1994 ISBN 1-899044-06-X

Stanley, P., **Robert Bakewell and the Longhorn Breed of Cattle**, Farming Press, Ipswich, 1995

Stout, A., **The Old Gloucester 'Story of a Cattle Breed'**, Alan Sutton, Gloucester, 1980

Alderson, L., **A Breed of Distinction**, Countrywide Livestock, (*White Park cattle*)

Hall, S.J. & Clutton-Brock, J., **Two Hundred Years of British Farm Livestock**, HMSO, Natural History Museum, London, 1995

Porter, V., Cattle, **A Handbook to the Breeds of the World**, Christopher Helm, 1991

Dobie, F.J., **The Longhorns**, Nicholson & Watson, London, 1943 (*Texas Longhorns; more recent editions are available.*)

Turner, F. N., **Herdsmanship**, Faber & Faber, London, 1952 (*This is a 'natural husbandry' book from the 1950s, some odd ideas but full of wisdom, all Newman Turner titles are worthy of your attention.*)

Heath-Agnew, E., **The History of Hereford Cattle and their Breeders**, Duckworth, London, 1983

Some of the above titles are out of print, try your library or the Internet.

Cookery

Its all very well producing the finest beef, but it helps to know how to make the best possible use of it. The following cooks and chefs have led the way in promoting the finest quality meats from traditional breeds, all their books are worthy of your attention, but particularly,

Grigson, S., **Sophie Grigson's Meat Course**, Network (an imprint of BBC Books), date ISBN 0-563-37173-0

Various books by Anthony Worrall Thompson and Hugh Fearnley Whittingstall

www.farmingbooksandvideos.com